I0012923

Michael Jaeger

Self-Managing Publish/Subscribe Systems

Michael Jaeger

Self-Managing Publish/Subscribe Systems

Foundations, Algorithms, and Analysis

VDM Verlag Dr. Müller

Impressum/Imprint (nur für Deutschland/ only for Germany)
Bibliografische Information der Deutschen Nationalbibliothek: Die Deutsche Nationalbibliothek
verzeichnet diese Publikation in der Deutschen Nationalbibliografie; detaillierte bibliografische
Daten sind im Internet über http://dnb.d-nb.de abrufbar.
Alle in diesem Buch genannten Marken und Produktnamen unterliegen warenzeichen-, marken-
oder patentrechtlichem Schutz bzw. sind Warenzeichen oder eingetragene Warenzeichen der
jeweiligen Inhaber. Die Wiedergabe von Marken, Produktnamen, Gebrauchsnamen,
Handelsnamen, Warenbezeichnungen u.s.w. in diesem Werk berechtigt auch ohne besondere
Kennzeichnung nicht zu der Annahme, dass solche Namen im Sinne der Warenzeichen- und
Markenschutzgesetzgebung als frei zu betrachten wären und daher von jedermann benutzt
werden dürften.

Coverbild: www.purestockx.com

Verlag: VDM Verlag Dr. Müller Aktiengesellschaft & Co. KG
Dudweiler Landstr. 125 a, 66123 Saarbrücken, Deutschland
Telefon +49 681 9100-698, Telefax +49 681 9100-988, Email: info@vdm-verlag.de
Zugl.: Berlin, TU, Diss., 2007

Herstellung in Deutschland:
Schaltungsdienst Lange o.H.G., Zehrensdorfer Str. 11, D-12277 Berlin
Books on Demand GmbH, Gutenbergring 53, D-22848 Norderstedt
Reha GmbH, Dudweiler Landstr. 99, D- 66123 Saarbrücken
ISBN: 978-3-639-07363-8

Imprint (only for USA, GB)
Bibliographic information published by the Deutsche Nationalbibliothek: The Deutsche
Nationalbibliothek lists this publication in the Deutsche Nationalbibliografie; detailed
bibliographic data are available in the Internet at http://dnb.d-nb.de.
Any brand names and product names mentioned in this book are subject to trademark, brand or
patent protection and are trademarks or registered trademarks of their respective holders. The use
of brand names, product names, common names, trade names, product descriptions etc. even
without
a particular marking in this works is in no way to be construed to mean that such names may be
regarded as unrestricted in respect of trademark and brand protection legislation and could thus
be used by anyone.

Cover image: www.purestockx.com

Publisher:
VDM Verlag Dr. Müller Aktiengesellschaft & Co. KG
Dudweiler Landstr. 125 a, 66123 Saarbrücken, Germany
Phone +49 681 9100-698, Fax +49 681 9100-988, Email: info@vdm-verlag.de

Copyright © 2008 VDM Verlag Dr. Müller Aktiengesellschaft & Co. KG and licensors
All rights reserved. Saarbrücken 2008

Produced in USA and UK by:
Lightning Source Inc., 1246 Heil Quaker Blvd., La Vergne, TN 37086, USA
Lightning Source UK Ltd., Chapter House, Pitfield, Kiln Farm, Milton Keynes, MK11 3LW, GB
BookSurge, 7290 B. Investment Drive, North Charleston, SC 29418, USA
ISBN: 978-3-639-07363-8

Preface

Acknowledgements

This book would not have been possible without the support of various people whom I would like to thank here. First of all, I thank my advisors Prof. Dr. Hans-Ulrich Heiß and Prof. Dr. Kurt Geihs for encouraging me to start this Ph.D. project and supporting me during the last four years. I am very grateful to PD Dr. Gero Mühl for his great support, many inspirations, the countless challenging discussions, and the encouragement in desperate moments (there were quite a few of them). Moreover, I thank my other dear colleagues Helge Parzyjegla, Dr. Klaus Herrmann, and Prof. Dr. Matthias Werner for their good advices, their support, and the fruitful cooperation. My work was also supported by *Deutsche Telekom Stiftung* with a generous fellowship. I highly appreciate this support which allowed me to concentrate on pursuing my research.

There were various other people involved in the non-technical development of this book. The most important are certainly my parents who always supported me in my decisions and gave me the backing to finally complete this big project. Nothing would (obviously) have been possible without them. I also thank my friends all over the world who supported me and endured all my mood swings which came along with this work: Dr. Jana Binder, Babette Brühl, Dr. Joaquín García-Alfaro, Verena Kowald, Jacqueline Laux, Dr. Michael Maschke, Dr. Kilian Planck, and Marc Träbert, just to name a few.

i

Publications

Parts of this book are based on papers published during the work on my Ph.D. thesis. Chapter 2 is partly based on joint work with Klaus Herrmann, Gero Mühl, Helge Parzyjegla, and Matthias Werner [118]. Chapter 3 is based on joint work with Ludger Fiege, Klaus Herrmann, Gero Mühl, Andreas Ulbrich, and Torben Weis [82, 117] and parts that have been published in [116, Chapter 8.2.4]. Chapters 4 and 5 are based on [80, 81] and on joint work with Klaus Herrmann, Gero Mühl, Helge Parzyjegla, and Matthias Werner [83, 84, 122].

Contents

Contents

List of Figures

List of Figures

x

List of Tables

List of Symbols

a Link to be inserted into the broker overlay topology, page 103

$\mathcal{A}$ Generic wrapper algorithm used to render an arbitrary routing algorithm $\mathcal{R}$ self-stabilizing, page 53

B A broker, page 15

b Bandwidth saved when applying self-stabilizing identity-based routing instead of flooding, page 68

b' Bandwidth saved when applying self-stabilizing identity-based routing instead of flooding in an arbitrary broker overlay network, page 90

$\mathcal{B}$ Set of all brokers in the system, page 16

$\mathfrak{B}$ Behavior of a system $\mathfrak{S}$, page 21

b_c Bandwidth needed for control traffic when applying self-stabilizing identity-based routing instead of flooding, page 51

b_f Bandwidth needed for (self-stabilizing) flooding in content-based routing, page 67

b_l Bandwidth needed for leasing traffic when applying self-stabilizing identity-based routing instead of flooding, page 72

b_n Bandwidth needed for notifications in self-stabilizing identity-based routing, page 50

b_p Expected additional bandwidth when using self-stabilizing identity-based routing, page 78

b_s Bandwidth saved in notification forwarding when applying identity-based routing instead of flooding, page 70

C Child broker, page 84

$\mathcal{C}(B)$ Set of child brokers of broker B, page 84

$c(f_M)$ Set of constituting filters that have been merged to the filter f_M, page 45

$C_{i,j}$ Cost of an indirect connection between two brokers B_i and B_j based on communication and processing costs taking common traffic into account, page 205

$\mathfrak{C}$ Control input of a system, page 21

$\text{cost}(n, T)$ Total cost for distributing notification n in the broker overlay tree T, page 188

δ Delay of communication links, page 37

d Diameter of the network, page 37

Δ Stabilization time, page 38

D Distance between two spanning trees $\mathfrak{T}_1$ and $\mathfrak{T}_2$, i.e., the number of edges that are contained in $\mathfrak{T}_2$ but not in $\mathfrak{T}_1$, page 106

Δ_a Stabilization time of the self-stabilizing routing stack which uses advertisements and the generic wrapper algorithm, page 60

Δt Time period between two consecutive broadcast messages sent by a broker, page 208

Δ_g Stabilization time of the generic wrapper algorithm, page 55

$\Delta_{i'}$ Stabilization time of idempotent self-stabilizing identity-based routing, page 42

Δ_r Stabilization time of the reconfigurable self-stabilizing publish/subscribe system, page 175

Δ_s Stabilization time of self-stabilizing simple routing, page 40

D_T^n Delivery tree for notification n in tree T, page 188

η Maximum number of hops neighbor brokers are away from each other, page 199

f A filter, page 14

$\mathcal{F}$ Set of all filter classes, page 85

$I(\mathcal{S})$ Number of identical notifications all brokers stored in the set $\mathcal{S}$ consume ($I_{i,j}$ denotes the common interest of brokers B_i and B_j from the perspective of B_i), page 203

i Number of routing entries for identical filters in identity-based routing, page 41

$\mathfrak{I}$ Input of a system, page 21

$\mathcal{I}$ Wrapper algorithm for rendering a correct routing algorithm robust against repeated (un)subscriptions, page 58

k Number of levels of the complete broker overlay network tree, page 62

L_B List of local clients of broker B, page 16

l Number of communication links in a complete broker overlay tree network, page 63

l' Number of communication links in an arbitrary broker overlay network, page 87

λ Interarrival rate between two consecutive subscriptions, page 49

m^{b} The begin message marks the start of the reconfiguration process, page 136

m^{e} An end message marks the end of the reconfiguration on one side of the reconfiguration cycle, page 141

m^{l} A lock message is used to lock the reconfiguration cycle prior to starting the reconfiguration, page 134

m^{s} The separator message is used to separate gray from black messages during reconfiguration, page 140

m^{r} Stores routing entries that have to be moved from the brokers on the old link to the brokers on the new link during the reconfiguration, page 136

m_o Toggle rate of a broker in self-stabilizing identity-based routing, page 76

$m_o^f(B_l)$ Toggle rate of a leaf broker B_l with self-stabilizing identity-based routing for one filter class f, page 75

m Degree of the complete broker overlay network tree, page 62

m^k The unlock message frees the brokers on the reconfiguration cycle for the next reconfiguration after the current reconfiguration is ended, page 143

m_B^K Broadcast message containing local information a broker B periodically sends to the brokers in its neighborhood $\mathcal{N}_\eta$, page 199

μ Rate at which subscriptions are removed from the system, page 64

μ^{-1} Expected lifetime of a subscription with exponentially distributed lifetimes, page 49

$\bar{\mathfrak{N}}$ Expected number of active subscriptions in the system, page 66

n A notification, page 14

$N(f)$ Set of notifications a filter f matches, page 14

N_B List of neighbor brokers of broker B, page 16

$\mathcal{N}_L$ Total number of leaf brokers in the broker overlay network, page 63

$\mathcal{N}_\eta(B)$ Set of brokers in the η-hop neighborhood of broker B in the broker overlay network, page 199

$\mathcal{N}_s$ Number of active subscriptions in the system, page 49

$\mathfrak{n}$ Number of subscriptions per leaf and filter class, page 66

$\bar{\mathfrak{n}}$ Expected number of subscriptions per leaf and filter class, page 66

o Expected number of control messages per toggle in self-stabilizing identity-based routing, page 74

$\mathfrak{O}$ Output of a system, page 21

ω Rate at which exponentially distributed notifications are published, page 64

ω^{-1} Expected delay between two consecutive publications with exponentially distributed publications, page 49

$\mathcal{P}(B)$ Set of brokers on the path from broker B to the root broker (exclusive), page 86

$p_0^f(B)$ Probability that a given broker B has no local subscription for one filter class f, page 83

p_i Probability that a broker has i subscriptions for one filter class, page 66

$p_l(i,j)$ Probability of choosing broker B_j as the subscriber hosting broker for a job, where the publisher is connected to B_i based on the distance between B_i and B_j, page 236

$p_s(i,j)$ Probability of choosing broker B_j as the subscriber hosting broker for a job, where the publisher is connected to B_i based on a combination of B_i's load value and the distance between B_i and B_j according to the parameter locality, page 236

$P_0^f(B)$ Probability that a given broker B is in state 0 for one filter class f, page 84

π Leasing period for the second-chance algorithm, page 36

$P(n)$ Broker with the local client that published notification n, page 185

P_{sub} Set of subscriptions that have to be moved from broker B_1^r (B_2^r) to B_1^a (B_2^a) during reconfiguration, page 132

Q_{not} Queue of the brokers connected to a during reconfiguration that is needed to ensure FIFO or causal notification order, page 129

Q_{sub} Queue of the brokers connected to a during reconfiguration that is needed to ensure FIFO (un)subscription order, page 128

r Link to be replaced by a in the broker overlay topology, page 103

ρ Refresh period of the subscribers, page 37

$\mathcal{R}$ Correct routing algorithm, page 53

$\mathfrak{R}$ Regular input of a system, page 21

$\mathcal{S}$ Set of brokers, page 203

s A subscription, page 14

$\mathfrak{S}$ A system defined by input, output, and behavior, page 21

$\mathcal{S}(n)$ Set of subscriber-hosting brokers that subscribed for a filter matching n, page 185

T_B Routing table of broker B, page 16

$\mathfrak{T}$ A broker overlay spanning tree, page 106

$V(n)$ Set of brokers with clients connected that subscribed for n or published n, page 188

$\mathfrak{W}$ Performance criterion, page 21

x^f Number of occupied remote routing entries of the subtree rooted in broker B for one filter class f, page 84

y Portion of notification traffic saved when applying self-stabilizing identity-based routing instead of flooding, page 69

z Number of filter classes, page 64

1 Introduction

1.1 Motivation

Over the last decades, computer systems of every scale have enriched our lives ranging from the World Wide Web on the one end to MP3 players and cell phones on the other end. The availability of wireless networking standards and the cheap price of the hardware needed to build computer networks has pushed the proliferation of computer and networking technology further into our day-to-day lives. With every year, we are getting closer to the vision of Internet access anywhere at anytime for affordable access fees. Moreover, the "Internet of things"[1] as the ultimate vision of an interconnected world gets within reach. This technical evolution fosters the development of new applications that take *social* as well as *local* (in terms of geography) aspects into account. Moreover, the current development shows that not only the number of *consumers* of content in those systems grows, but also that of *producers*. This development is demonstrated by new applications gaining growing prominence in the recent past like weblogs, wikis, and social networks like FACEBOOK[2], just to name a few. The vast sources of information in existence and the desire to combine them in order to obtain an added value drive the classical distributed programming paradigm based on the remote procedure call to its limit.

A good example which is often used for aggregating information and which also reflects the problems of large scale information dissemination systems based on

---

[1]http://www.itu.int/internetofthings (last visit: 2007-10-04)
[2]http://www.facebook.com (last access: 2007-10-03)

synchronous mechanisms is *Rich Site Summary*[3] (*RSS*). RSS is a family of data format specifications for providing summaries of website contents. It is mostly used to provide short information snippets about recent changes on a website which can be aggregated by other websites or subscribed for and read by users with the help of RSS reader software. The basic mechanism to get informed about updates relies on polling. In the past, this led to severe problems on some web servers hosting popular websites because polling can easily be automated in the client software and a small update interval results in a high timeliness. Since many users nowadays possess a broadband Internet connection and are interested in up-to-date information, a low polling period is often chosen which can overload the servers similar to a denial-of-service attack. This problem could be solved by relying on an active *push* mechanism which proactively delivers news to the clients in contrast to a polling-based scheme, because the update rates of most RSS sources are rather low.

Event-Based Computing

Today, we are already in the very comfortable situation that we can access a vast range of information through the Internet. This enables us to track changes happening in the world and react to them accordingly. However, it is not feasible for humans to manually track changes from a large set of sources—and a respective automated management sooner or later reaches system limits as described for RSS above. This insight evolved to the idea of *event-based computing* which is easy to understand because it resembles our daily life much more than the classical synchronous procedure call. Similar to getting informed when our CD-player is repaired, we want to get informed when a website or stock quote changes instead of permanently asking for the current state (which resembles a *polling* model). Actions are, thus, the consequence of an *event*, which indicates an arbitrary state change.

---

[3]Also called RDF Site Summary or Really Simple Syndication.

The paradigm of event-based programming has, for example, been successfully applied to graphical user interface programming with JAVA SWING and *event-condition-action rules* (ECA-rules) in the area of active databases [123]. The idea of using events as the central programming abstraction, however, is still not widespread although it has already found its way into teaching at universities [72]. The event paradigm is especially well suited to data-centric dynamic communication like workflow systems, logistics, and systems monitoring (e.g., intrusion detection). It increases the flexibility and fosters the decoupling of components. The Enterprise Service Bus (ESB) [41] is just one prominent example, where the event-based computing paradigm is used to integrate enterprise applications via loose coupling.

For event-driven (distributed) systems, it is common to rely on a (distributed) mediator which is responsible for delivering events from producers to consumers. This role can be taken over by a database, a tuple space, or a publish/subscribe middleware, for example. The relevance of event-driven approaches is accommodated by a respective support mechanism in various implementations of popular middleware standards like the CORBA EVENT AND NOTIFICATION SERVICE, JAVA MESSAGE SERVICE (JMS), and the DATA DISTRIBUTION SERVICE FOR REAL-TIME SYSTEMS (DDS).

Towards Self-Managing Publish/Subscribe Systems

Publish/subscribe middleware is an excellent choice for the implementation of event-based systems which are well suited to dynamic environments. This is mainly due to the simple, yet elegant, interface of publish/subscribe systems which decouples the clients in time, space, and program flow [61]. The requirements posed on the publish/subscribe middleware by dynamic environments, however, demand for new approaches regarding its management.

The management of publish/subscribe systems comprises a wide variety of tasks including reconfigurations of the broker overlay network topology (e.g., for maintenance or performance optimization), adding (removing) brokers to (from) the system (e.g., to increase the performance of the system), resetting brokers in case of faults to bring the system back into a working state, and adapting publication rates of publishers (e.g., to avoid congestion). It can be assumed that the systems that are built on top of publish/subscribe middleware will significantly grow in the near future. Large-scale software systems, however, often turn out to come along with not only quantitative but also qualitative new challenges regarding the management due to various reasons explained in the following.

Growing Management Costs. Thanks to significant advances in research on software engineering in the last decades, it is today feasible to engineer and build software systems of considerable scale. Running and maintaining large-scale software systems, however, is an expensive task because it often requires large numbers of well-trained personnel. As a consequence, the total cost of ownership of large systems is increasingly dominated by *administration costs*.

Exploding Complexity. Experience shows that the *complexity* of large-scale systems is the key issue which makes manual management costly if not impossible. Research on automated management is, thus, gaining increasing interest in academia and industry. This development is also driven by the growing demand for interconnecting systems to form new systems providing value-added services. As a consequence, the complexity of the resulting systems grows, as well as the complexity of the ways in which those systems fail [139]. In such a setting, human operators are no more able to keep the overview of the system. They are, thus, unable to draw management decisions because the consequences of their actions are unforeseeable.

Unmanageable Dynamics. Besides the complexity of large-scale systems, the *dynamics* of modern computer systems pose another challenge. In the past, most systems were deployed in a rather static environment, for example, inside a company. The trend towards opening systems to the Internet and fostering interconnecting them has significantly increased the dynamics those systems are exposed to and requires fast reactions from the system administrators. Paired with the growing complexity, the task of manually managing such a system becomes extremely difficult if not infeasible. The growing popularity of mobile devices and their integration into existing systems also push this development.

Due to the above reasons it is, thus, of significant importance to enrich publish/subscribe systems with features which enable them to manage themselves minimizing human intervention thereby. This would not only facilitate running complex systems in face of dynamic changes but would also save costs. Automated management includes fault management and performance optimizations which are both extremely important in dynamic scenarios. Regarding fault management, recovery guarantees are getting increasingly important since the probability of faults grows with the system size. In order to support dynamic environments, reconfigurations need to be carried out. It is, thus, important to support them at runtime while not interrupting the system service. The next step is to enable the system to draw reconfiguration decisions by itself in order to improve its performance. The ultimate goal is to build a system which does not need any external management and is able to run completely autonomously.

1.2 Shortcomings of Current Approaches

In a publish/subscribe system, the task of disseminating events from producers to consumers is taken over by a notification service. A distributed implementation of the notification service often consists of a set of brokers which are interconnected to form an overlay network. Research in the area of publish/subscribe systems has been very agile in the last few years. However, the task of managing and reconfiguring publish/subscribe systems at runtime has rarely been researched yet, particularly in the context of automation. The shortcomings of current approaches are listed in the following.

Lack of Recovery Guarantees in the Face of Faults. Fault management is of significant importance in dynamic environments. Most research in the past has not dealt with faults at all, or has tried to mask occurring faults. Another common approach is to layer publish/subscribe systems on top of peer-to-peer routing substrates, thereby inheriting their fault resilience properties on the overlay network layer [127]. However, those systems are only designed with particular faults in mind, like link or broker failures neglecting the higher application layers like perturbed routing table entries, for example. The increasing system complexity and miniaturization, however, raises the probability of faults which may lead to arbitrary corruptions of data structures. This issue has often been neglected in fault tolerance research for publish/subscribe systems in the past to such an effect that recovery guarantees in case of transient faults were not provided.

Service Interruption Due to Reconfigurations. Reconfiguration is an important topic for systems exposed to dynamic changes. In the case of publish/subscribe middleware it may be necessary to reconfigure the broker overlay topology for maintenance or performance optimization. In both cases, it is required that the reconfiguration is seamless and transparent for the users and that the system service is not interrupted. Algorithms proposed in the past for topology reconfiguration do not prevent message loss due to reconfigurations and do not guarantee a particular message ordering.

Only Manual Adaptation. Adapting a system automatically to changes in dynamic environments becomes particularly interesting because the sheer size of Internet-wide systems makes it expensive or even impossible to manually manage them. A similar situation arises for the new generation of small-scale networks which are deployed in isolation (e.g., sensor networks) or are operated by non-experts who are unable to manage them (e.g., home consumer networks). While peer-to-peer substrates are able to cope with changes and can even adapt to the underlying physical network infrastructure, they do not adapt to the application behavior on the higher publish/subscribe layers yet. Research on publish/subscribe in the past has either assumed a static broker network or proposed heuristics which do not take network *and* application dynamics into account.

Lack of Analytical Models. It is common practice to evaluate publish/subscribe systems by simulation. However, there is no standard simulation environment available up to now and the impact of parameter variations on the system behavior is often unclear. Analytical models can serve to create a deeper understanding of publish/subscribe systems and may furthermore relieve researchers from expensive simulations.

1.3 Focus and Contribution

The focus in this book is on fault management, dynamic reconfigurations, and performance optimizations as well as formal analysis. It concentrates on four topics which are detailed in the following.

Guaranteed Recovery After Transient Faults. Fault masking is expensive and requires that all faults are specified in advance. Moreover, if a fault occurs which cannot be masked, the system can move into an invalid state from which it may not recover—even if the fault was only transient. Self-stabilization provides valuable recovery guarantees in dealing with faults. This book presents algorithms for selected routing algorithms and also a general wrapper algorithm which renders arbitrary correct routing algorithms self-stabilizing. The overhead induced by self-stabilizing mechanisms is evaluated in a simulation study. Following an approach based on self-stabilization it is possible to provide recover guarantees in the face of transient faults.

Seamless Broker Overlay Topology Reconfiguration. One of the most important management tasks in a publish/subscribe system is to reconfigure its topology.

Reconfigurations are needed to recover from faults, to optimize system performance, and to deal with joining and leaving brokers. Solutions presented in the literature either conflict with the message completeness requirement or with the ordering imposed on notification delivery and control messages or with both of them. This book presents algorithms for regular and self-stabilizing publish/subscribe systems which permit a seamless reconfiguration of the broker overlay topology at runtime, preventing message loss and providing message ordering if required.

Self-Optimizing Broker Overlay Topology. Most distributed publish/subscribe systems are realized by a broker overlay network, where the brokers cooperatively provide the event notification service. The structure of this overlay network has an impact on notification forwarding and, thus, the performance of the notification service. The problem of finding a "good" broker overlay topology with respect to communication *and* processing costs is analyzed in this book. As a result, it will become clear that the problem of finding an optimal broker overlay topology is NP-hard. To find a good solution anyway, a distributed heuristic which renders the broker overlay network self-optimizing is presented. This way the system is enabled to autonomously adapt to changes in the physical network as well as to changes on the application layer.

Formal Analysis. A formal analysis is often the key to a deeper understanding of systems. In publish/subscribe research, analytical models are still rare. This has led to the situation that most work is evaluated in simulation studies which are time consuming and often hard to compare. In this book, a formal analysis based on probability theory is presented which represents a first step to analyzing publish/subscribe systems without the need for extensive simulations. The analysis is carried out as part of the work on self-stabilizing content-based routing.

1.4 Methodology

The scenarios targeted by this book comprise large-scale as well as small-scale networks. In order to evaluate the algorithms presented, a simulation-based approach has been chosen, and an analysis where appropriate. The reasons for relying on simulations are manifold. Running large-scale experiments requires an adequate infrastructure which is ideally under the sole control of the experimenter. Running experiments on a scale bigger than 10 computers becomes difficult since an adequate infrastructure is hard to obtain. Even if the infrastructure is available, faults which are independent of the algorithm studied may tamper the results. Moreover, these experiments cannot be parallelized. As a consequence, repetitions of the experiments (in order to obtain statistically relevant results) have to be serialized which increases the time needed. Additionally, network dynamics prevent an exact reproduction of the results.

For the simulation studies, a new simulation platform has been implemented which provides the opportunity to replay experiments in order to verify the results. Since the experiments explicitly target the higher layers of the algorithm stack, existing simulation platforms like NS2 have not been used. Evaluating the algorithms with "real" applications on platforms like PLANETLAB would be an interesting next step, although some of the problems discussed above will remain.

Besides simulation studies, formal methods are used to analyze the proposed algorithms for self-stabilizing content-based routing. The approach builds on Markov chains and probability theory in order to evaluate publish/subscribe systems applying content-based routing and calculate the overhead induced by the self-stabilizing algorithms presented. The results thus gained confirm the results from the simulations carried out.

1.5 Organization of this Book

The topics and the organization of this book are depicted in Figure 1.1. It starts with a basic introduction to publish/subscribe systems, the notion of self-management including self-optimization and self-stabilization in Chapter 2 which closes with the basic assumptions this book builds upon.

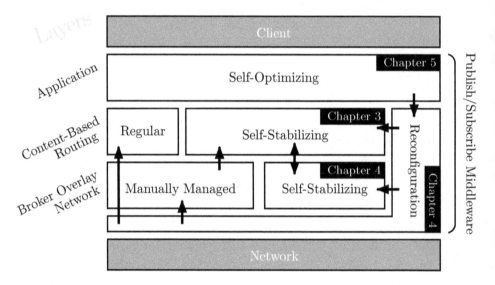

Figure 1.1: Topics and organization of this book in the algorithm layers

In the first part of Chapter 3, algorithms which render content-based routing in publish/subscribe systems self-stabilizing are presented. This comprises algorithms for selected routing algorithms and also a general wrapper algorithm which renders arbitrary correct routing algorithms self-stabilizing. The overhead induced by self-stabilization is evaluated in a simulation study and its performance is compared to flooding. In the second part of Chapter 3, a stochastic analysis of the message complexity is presented which also allows to analyze the overhead induced by self-stabilization without performing extensive simulations.

Chapter 4 deals with reconfiguring the broker overlay network. In the first part of this chapter, an algorithm for regular publish/subscribe systems (i.e., those which are not self-stabilizing) is presented. In the second part, a reconfiguration algorithm for self-stabilizing publish/subscribe systems is presented. To achieve this, results presented in Chapter 3 are incorporated. Both algorithms prevent message loss and are able to preserve FIFO-publisher and causal ordering.

In Chapter 5, the problem of finding an optimal topology for the broker overlay network is analyzed. Then, a heuristic is presented which adapts the broker topology according to the message flows in a decentralized manner. Its performance is compared in an extensive simulation study with that of two other heuristics proposed in literature. The reconfigurations can, for example, be carried out using the algorithms proposed in Chapter 4.

Chapter 6 summarizes the results and presents the conclusions. Furthermore, problems which remain open are discussed and areas for future work are sketched.

1 Introduction

2 Basics and Model

Contents

2.1 Introduction

In this chapter, the foundations for this book are laid. First, the publish/subscribe system model and architecture is introduced (Section 2.2). Then, the notion of self-management and self-managing publish/subscribe systems used in this book is explained in Section 2.3. The notion of self-optimization is discussed in Section 2.4, where its relation to self-management is discussed, too. Section 2.5 gives an introduction to self-stabilization and its relationship to self-management. This chapter closes with Section 2.6, where the basic assumptions underlying this book are listed and explained.

2.2 Publish/Subscribe

The publish/subscribe communication paradigm is event-based and belongs to the general class of group communication paradigms. In the following, the publish/subscribe system model and architecture is introduced which is used in this book. Routing and matching are central issues in publish/subscribe systems and various algorithms have been proposed for them. A short introduction to selected algorithms for both topics will be given in this section.

2.2.1 System Model

A publish/subscribe system consists of a *notification service* and a set of clients that interact via a notification service. Clients can act as *publishers* (also: *producers*) or *subscribers* (also: *consumers*). While the former publish *notifications*, the latter issue *subscriptions* that contain *filters* that match the set of notifications the subscriber is interested in (on the implementation level, subscriptions may contain more information than just the filter like additional meta-data, for example). The set of notifications matched by filter f is given by $N(f)$. The notification service is responsible for routing published notifications to those clients which issued a subscription with a filter which matches these notifications. This content-based addressing scheme is sometimes viewed as the essential difference between "channel-based" schemes like IP multicast and publish/subscribe [34]. Another essential feature of publish/subscribe systems is the potential decoupling of the clients in space, time, and program flow [61].

The basic interface of the notification service consists of the following operations which can be used by the clients: pub(n) for publishing a notification n and sub(s) (and unsub(s)) for (un)subscribing for a subscription s. The notification service calls the operation notify(n) on a client in order to inform it about a new notification n it subscribed for. The semantics of the (un)subscribe operation are the following. The notification service manages the set of subscriptions $\mathbb{S}$ for all clients. If a client subscribes to a new subscription s, this subscription is added to $\mathbb{S}$:

$$\mathsf{sub}(s) \Rightarrow \mathbb{S} \leftarrow \mathbb{S} \cup \{s\} \qquad (2.1)$$

In case a client unsubscribes for a subscription for which it is not currently subscribed, this operations has, thus, no effect. Similarly, issuing a subscription for a filter the client is already subscribed to, has no effect, too.

Analogous, an unsubscription for a subscription s removes the subscription from $\mathbb{S}$:

$$\mathsf{unsub}(s) \Rightarrow \mathbb{S} \leftarrow \mathbb{S} \setminus \{s\} \qquad (2.2)$$

Sometimes, the notification service also supports the operations $\mathsf{adv}(f)$ and $\mathsf{unadv}(f)$ which are used to announce or recall *advertisements*. They are issued by publishers to announce which kind of notifications they are going to publish. The semantics of the operations are analogous to those of the respective operations for subscriptions. Similar to subscriptions, advertisements contain filters and are used to optimize routing inside the notification service.

Figure 2.1 illustrates the system model and the operations described above.

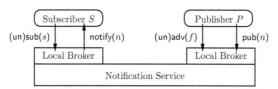

Figure 2.1: Publish/subscribe system model

2.2.2 System Architecture

This book focuses on a distributed implementation of the notification service which consists of a set of *brokers* forming a *broker overlay network*. An acyclic topology for the broker overlay network is required which is a prerequisite for using advanced routing algorithms such as covering-based routing. Moreover, FIFO links between any two brokers are required; this is common in literature because it simplifies maintaining message ordering and preventing message duplicates (regarding both control messages and notifications). The set of all brokers in the system is called $\mathcal{B}$. Figure 2.2 illustrates the architecture of a publish/subscribe system referred to in this book.

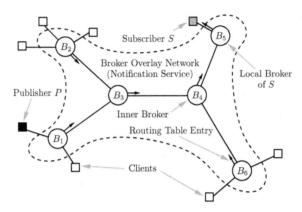

Figure 2.2: Publish/subscribe system architecture

Each broker B maintains a routing table T_B in order to decide whether and if where to forward an incoming notification to. A routing table holds *routing entries* which are ⟨filter, destination⟩ pairs, where a *destination* is a neighbor broker or a local client. In Figure 2.2, the routing table entries are depicted as little arrows. It is differentiated between *local* and *remote* routing entries (in Figure 2.2, all routing entries are remote except for the one at B_5). N_B is the set of remote destinations of B (e.g., $N_{B_5} = \{B_4\}$ in the example figure) and L_B is the set of local destinations of B (e.g., $L_{B_5} = \{S\}$).

2.2.3 Matching, Forwarding, and Routing

Matching

When a broker B receives a notification n, it decides based on its routing entries whether to forward it to a destination or not. Therefore, B checks whether a filter for a destination matches n. The problem of determining all filters which match a notification is termed the *matching problem*. With channel-based matching each notification is tagged with a selected channel and subscribers can only subscribe to channels. Subject-based matching is more expressive, because notifications are published with respect to a selected subject, where the subjects themselves can be arranged in a hierarchical fashion in contrast to "flat" channels. The most expressive matching relies on the content of the notification. Many different algorithms have been proposed in the past for content-based matching, most notably the counting algorithm [160], decision trees [6], and binary decision diagrams [29].

Forwarding

Mühl et al. distinguish between matching and *forwarding* [116]. While matching refers to the problem of determining all filters in a routing table which match a given notification, the forwarding problem consists of determining all destinations a given notification must be sent to. The difference between both problems is that for notification forwarding it may not be necessary to find all matching filters in a routing table for a given notification.

Routing

Besides the fundamental matching and forwarding problems, *notification routing* is also an important issue. The routing problem deals with routing notifications from publishers to subscribers through the broker network. Therefore, routing tables are maintained by the routing algorithm. The simplest routing algorithm is *flooding*, where every notification is sent to all brokers in the system. The brokers then match each received notification against their local subscriptions to decide whether to deliver the notification to their local clients or not. It is obvious that this routing strategy is suboptimal if there is a significant number of brokers not having a local client that is interested in them. A more sophisticated approach is *simple routing*, where subscriptions are flooded in the broker network. On receiving a subscription, the receiving broker installs a routing entry in its routing table pointing to the broker from which it received the subscription. In this case, each subscription is treated in isolation from other subscriptions and notifications are routed between brokers according to the routing table entries.

Advanced Routing Algorithms. More sophisticated routing algorithms try to exploit similarities between the filters stored in subscriptions and are often referred to as *advanced routing algorithms*. In the case of *identity-based routing*, subscriptions are not forwarded to a neighbor broker if an identical subscription has already been sent there in the past which is still active. Two subscriptions or filters f_1 and f_2 are identical if the set of notifications both match are identical (i.e., $N(f_1) = N(f_2)$). With *covering-based routing*, subscriptions are tested with respect to their covering relation. A filter f_1 covers a filter f_2 if the set of notifications it matches is a superset of the set of notifications f_2 matches (i.e., $N(f_1) \supseteq N(f_2)$). In this case, a subscription s does not need to be forwarded to a neighbor broker if a covering subscription has been sent there in the past. With *merging-based routing*, brokers try to merge filters in order to create a new filter. This new filter either perfectly matches the set of notifications covered by the constituent filters (*perfect merging*) or it covers a superset of those notifications (*imperfect merging*). The new filter that results from merging can accordingly be used to perform subscription forwarding optimizations similar to covering-based routing.

Hierarchical vs. Peer-to-Peer Routing

With the help of advanced routing algorithms it is often possible to reduce the traffic caused by control message forwarding. Additionally, the routing table sizes decrease which can lower the matching overhead. Regarding the dissemination of subscriptions in the broker overlay network it is distinguished between *hierarchical routing* and *peer-to-peer routing*. With hierarchical routing, every subscription is sent only towards a dedicated root broker R which does not forward them any further. Additionally, all notifications published are also forwarded at least to R. Figure 2.3(a) illustrates hierarchical routing in a scenario where B_6 subscribed for a filter which matches a notification n that is published at B_5. The gray arrows depict the path the notification is sent while the black arrows represent the routing table entries.

With peer-to-peer routing, there is no dedicated root broker. Thus, notifications do not need to be sent always to a dedicated broker but routing entries have to be installed in the whole broker network. Figure 2.3(b) illustrates peer-to-peer routing in the same setting used for the hierarchical routing example.

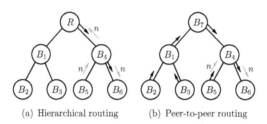

(a) Hierarchical routing   (b) Peer-to-peer routing

Figure 2.3: Routing options in publish/subscribe systems

Advertisements

While in the case of hierarchical routing subscriptions are only sent to a dedicated root broker (and notifications are sent there at least, accordingly), subscriptions may be flooded in the whole broker network with peer-to-peer routing. Advertisements can be used to build up routing tables for subscription forwarding just as regular routing tables are used for notification forwarding. The subscription routing tables ensure that subscriptions are only forwarded in directions, where brokers with publishers reside which may publish notifications that match this subscription. This way, the traffic caused by (un)subscriptions can be reduced if there are only a few publishers in the system producing notifications clients are interested in.

In Figure 2.4, an example scenario is depicted analogous to the one in Figure 2.3(b), where B_5 issued an advertisement which installs subscription routing entries in the whole broker network. Afterwards, B_6 issues a subscription s which matches a subset of the notifications the advertisement matches. Accordingly, s is only sent towards B_5 and the respective routing table entries are created on the path.

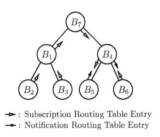

→ : Subscription Routing Table Entry
→ : Notification Routing Table Entry

Figure 2.4: Example for the application of advertisements

For advertisements, the same routing algorithms can be used as for subscriptions. A general comprehensive overview of publish/subscribe systems is given by Mühl et al. in [116].

2.3 Self-Management

The book at hand presents algorithms and concepts to render publish/subscribe systems self-managing. However, the meaning of *self-management* is rather fuzzy in the literature and often misses a formal definition besides its intuitive meaning. In this section, the notion of self-management based on the work published in [118] is introduced and put it into the context of publish/subscribe systems.

2.3.1 Basics

Intuitively, a system is self-managing with respect to a management goal, if it executes management tasks on its own without external (manual) intervention by a human operator in order to reach this goal. Although easy to understand, this description of self-management leaves several issues open which are tackled in the following definitions from [118].

Definitions

As a first step, the notion of a *system* is introduced which is based on a behavioral model by Willems and relies on input, output, and behavior [155].

Definition 1 (System). *A system $\mathfrak{S}$ is a tuple $\mathfrak{S} = (\mathfrak{I}, \mathfrak{O}, \mathfrak{B})$ with input interface $\mathfrak{I}$, output interface $\mathfrak{O}$, and behavior $\mathfrak{B}$.*

In analogy to Zadeh [163], a system $\mathfrak{S}$ is called *adaptive* if it performs acceptably well with respect to a performance criterion $\mathfrak{W}$.

Definition 2 (Adaptive System). *A system $\mathfrak{S} = (\mathfrak{I}, \mathfrak{O}, \mathfrak{B})$ is adaptive wrt. a set $I \subseteq \mathbb{I}$, a performance function p, and an acceptance criterion $\mathfrak{W}$ iff the following holds*

$$\forall i \in I : o = \mathfrak{B}(i) \Rightarrow p(i, o) \in \mathfrak{W}$$

Intuitively, a system is adaptive if it performs "well" in a range of scenarios characterized by a set of input functions I. For the definition of self-management it is necessary to differentiate the input fed into the system by $i \in I$ into *regular input* and *control input* (which "manages" the system) according to Lendaris [101]. In the following, $\mathfrak{R}$ stands for regular input and $\mathfrak{C}$ for control input. The input $\mathfrak{I}$ is the conjunction of regular and control traffic ($\mathfrak{I} = \mathfrak{R} \cup \mathfrak{C}$).

Definition 3 (Self-Manageable System). *A system $\mathfrak{S} = (\mathfrak{I}, \mathfrak{O}, \mathfrak{B})$ with $\mathfrak{I} = \mathfrak{R} \cup \mathfrak{C}$ is self-manageable wrt. I iff the following holds:*

1. *$\mathfrak{S}$ is adaptive wrt. I and*

2. *there exists a computable, non-anticipating behavior $C : (\mathbb{T} \rightarrow dom(\mathfrak{R}) \times dom(\mathfrak{O})) \rightarrow (\mathbb{T} \rightarrow dom(\mathfrak{C}))$ with:*

$$\forall i \in I \; \exists i' \in I \; : \; i' = i_{\mathfrak{R}} \circ C(i_{\mathfrak{C}}, \mathfrak{B}(i'))$$

A *self-manageable system* is, thus, an adaptive system for which the control input needed to make the system adaptive can be computed without requiring knowledge about the future (i.e., the function which computes the control input exhibits a non-anticipating behavior). Since control input $\mathfrak{C}$ is responsible for managing the system $\mathfrak{S}$, it can be determined if a system is *self-managing* by simply looking at the type of input and the output. If the system performs well and the input to the system contains no control input, then the system is self-managing.

Definition 4 (Self-Managing System). *A system $\mathfrak{S}$ is* self-managing *wrt. a set of input functions I iff it is adaptive wrt. I and $\mathfrak{C} = \emptyset$.*

It is important to note that self-management is strongly tied to the definition of adaptivity and its notion of performance. The term "self" refers to the ability to reach this adaptivity without external intervention.

Example

The definition of a self-manageable and self-managing system is illustrated in the following by the example of a simple room heating system. Assume that the user requires a comfortable room temperature between 22 °C and 25 °C. Thus, the performance function accepts all temperature curves which do not break these thresholds. Take the room together with the heating system as the system $\mathfrak{S}$. The input to the system comprises every external stimulation like the temperature outside, the temperature inside, and manual regulation of the heating system. Now, if the user controls the heating system to such an effect that he feels comfortable with the room temperature, the system is called adaptive because the output (the room temperature in this case) performs well since it meets the user's requirements.

Assume that it is possible to build an automaton which turns up or down the heating system according to the current room temperature in order to keep the room temperature within the given range. In this case, $\mathfrak{S}$ is self-manageable. This is certainly only the case if the outside temperature which has an influence on the room temperature is changing slowly enough and not too extreme (this limits the set of input functions I for which the system is adaptive or self-manageable). A new system $\mathfrak{S}'$ consisting of $\mathfrak{S}$ together with the automaton is self-managing with respect to the temperature requirements given by the user and the outside temperature constraints since it performs well without any external control input.

In the example, it is possible to test whether $\mathfrak{S}'$ is self-managing by looking at the input fed into the system and measuring the room temperature simultaneously. The system $\mathfrak{S}'$ is self-managing if the input does not contain control input and the room temperature is not lower than 22 °C and not higher than 25 °C.

2.3.2 Self-Management for Publish/Subscribe Systems

Important management tasks in the context of publish/subscribe systems are, for example, fault management, choosing an appropriate routing or matching algorithm, determining publication rates in order to not overload the system, and selecting the number of brokers in the system. Many of these decisions have to be drawn at design or deployment time. Enabling systems to be reconfigured or adapted at runtime requires further efforts.

In this book, the problems of providing recovery guarantees in face of faults and adapting the broker overlay topology to optimize the performance of the publish/subscribe system with respect to communication and processing costs are tackled. For automation of performance improvement, topological reconfigurations at runtime are integrated and the topology is reconfigured based on a distributed heuristic which takes notification flows into account and issues reconfiguration stimuli if beneficial. Testing the adaptivity of the resulting system can be accomplished, for example, by relying on the notion of c-adaptivity introduced by Herrmann in [75]. A system is c-adaptive if it performs within a constant factor c of the theoretical optimal performance. For a simulation study, Herrmann determined the factor c using the results of the experiments executed [75, Section 8.8.1].

2.4 Self-Optimization

The term self-optimization has been introduced as part of IBM's *Autonomic Computing Initiative* (ACI) and is meant to be used in the context of automated performance tuning [93]. It is used in the same lax meaning here.

Adapting the broker overlay topology perfectly fits the notion of *self-organization* as defined by Herrmann [75] because the *structure* of the system is changed (here: the broker overlay topology) in a decentralized fashion in order to improve its performance (or meet some performance requirements). However, the term *self-optimization* is used here since it better reflects the intention behind the development of the algorithms.

2.5 Self-Stabilization

The concept of *self-stabilization* was introduced by Dijkstra in 1974 and relies on the notion of *legitimate states* [56]. A system is in a legitimate state if its state conforms to a legitimate configuration with respect to its specification. According to Dijkstra, a system is self-stabilizing if and only if started in an arbitrary initial state, it is guaranteed that it eventually reaches a legitimate state. This means that if no fault happens, the system is guaranteed to stay in the set of legitimate states once it has reached one. These two properties have been termed *convergence* and *closure* by Arora and Gouda [11]. The maximum time needed for a system to converge back to a legitimate state is called its *stabilization time*.

The fault model of self-stabilization comprises arbitrary state perturbations which are temporary (*transient*). It is important to note that it is assumed that the program code itself as well as data stored in ROM will not be perturbed. Self-stabilization has been successfully used in several areas (e.g., for networking protocols like OSPF and RIP) and is very attractive for lowering management and implementation costs as almost no assumptions regarding failures are required. On the downside, no guarantees concerning the behavior of the system can be made while it is not in a legitimate state.

There are various ways to realize self-stabilization. Most solutions follow either the *detect and repair* or the *push* approach. Algorithms following the first try to detect faults and repair them accordingly. This approach can be used when the correct state of the system is independent of external input. A good example for an algorithm falling into this category is a self-stabilizing spanning tree algorithm, where each node detects whether its parent conforms to the definition of a legitimate state (e.g., in the algorithm by Afek, Kutten, and Yung [3]). Systems whose correct state relies on a soft state that depends on external input have to rely on a push approach, where the self-stabilizing algorithm constantly pushes the system into a correct state using external input. This is necessary because the system may be corrupted arbitrarily to such an effect that it is impossible to converge back to a correct state without information fed into the system from the outside.

Self-stabilization can be used to supplement or replace fault masking. This is of particular interest in large networked systems, where the probability of faults grows. The same is true for networks that consist of error-prone links and unreliable nodes like sensor and ad hoc networks. In Chapter 3, the issue of self-stabilizing content-based routing is discussed in greater depth and a definition of legitimate configurations of a publish/subscribe system is provided in order to define the set of legitimate states. The solution presented follows a push approach as described above because the correctness of the publish/subscribe system is defined at its interfaces and depends on the set of active subscriptions of each client which are assumed to reside outside the system. Thus, clients have to refresh their state regularly in order to facilitate convergence of the soft state in the system (i.e., the routing table contents) to a legitimate configuration.

According to the definition of self-managing systems given above, a self-stabilizing system is also self-managing with respect to the following inputs and set of performance functions. Any possible regular input is allowed which comprises also arbitrary faults. For the performance functions, every function is allowed that accepts only the correct behavior at least after the stabilization time when the last fault occurred. In this case, the system does not need any control input to work acceptably well.

Since an in-depth introduction to self-stabilization would go beyond the scope of this book, the interested reader may refer to the following literature on self-stabilization. Arora and Wang give a short introduction to self-stabilization, where they also discuss common misunderstandings [12]. A more complete and in-depth discussion is given by Dolev in his book on self-stabilization [57] and by Schneider in his survey of self-stabilization [138].

2.6 Basic Assumptions

In this book, some assumptions are made which might not hold in general. They are listed below and explained in the following.

Local Communication. The model assumes that the communication between clients and their local broker is free and instantaneous. This is, for example, the case if the clients and their local broker run on the same machine.

Cooperative Behavior. It is assumed that brokers behave in a cooperative manner, i.e., they act as defined by the routing algorithm and cooperate with the other brokers in forwarding messages. Thus, malicious brokers which might try to disturb the whole system are not taken into account.

Small Proportion of Control Traffic. In order to justify the application of content-based routing algorithms, it is assumed that the notification traffic is much higher than the control traffic caused by (un)subscriptions and (un)advertisements. If this does not hold, it would not be beneficial to employ a subscription- or advertisement-based approach and it could be favorable to fall back to simple flooding. Thus, for example only the notification traffic is considered in Chapter 5 when optimizing the broker overlay topology of a publish/subscribe system.

Moderate Dynamics. The control traffic in a publish/subscribe system is tied to the behavior of the clients which publish notifications and subscribe for them. As already discussed above, moderate dynamics are assumed in this case. Similarly, also moderate dynamics are assumed on the links and the broker nodes with respect to the communication and processing costs, respectively. This assumption will be discussed again in Chapter 5, where both costs are considered when optimizing the broker overlay topology. The same is true for faults which are tackled in Chapter 3 in the context of self-stabilization.

FIFO Links. It is required that the communication links in the broker overlay network exhibit a FIFO behavior. This property is important because it is this way guaranteed that messages cannot overtake each other in a static acyclic broker overlay network. Otherwise it could happen, for example, that an unsubscription overtakes a subscription, creating stale routing entries thereby. FIFO links are easy to implement and required by many publish/subscribe systems which employ a single acyclic broker overlay network. They are explicitly required in Chapters 3 and 4.

3 Self-Stabilizing Content-Based Routing

Contents

3.1 Introduction

Publish/subscribe systems are used to implement large-scale event-driven applications. The loosely-coupled style in the publish/subscribe communication paradigm has a lot of advantages as it clearly separates communication from computation. However, this separation poses severe challenges on handling faults—for system designers as well as for application developers. In the past, most research on fault tolerance in publish/subscribe systems has concentrated on fault masking in a restricted fault model that only regards link or node failures and only offers best-effort semantics. Corrupted or erroneously inserted messages as well as perturbed routing tables have not been considered at all so far. Fault masking is always specific to certain faults and the larger the system grows, the higher the probability gets that faults occur which cannot be masked. In this case, recovery is not guaranteed. Especially in large-scale scenarios, however, eventual recovery from transient faults is required.

Self-stabilization is a concept which fits well in this gap because it provides *guarantees* regarding the convergence to a correct state from any transient fault and also from permanent faults under certain conditions. The question may arise, whether the assumption that arbitrary faults or state corruptions of a system need to be considered. One argument against this approach may be that this is purely theoretical and may not happen in practice. Jayaram and Varghese showed that for asynchronous protocols and for nodes with non-volatile memory which can crash and restart with an initial state, any global state can be reached by solely dropping messages and crashing nodes which are common faults practice [85, 151]. Moreover, the probability of memory perturbations grows due to the miniaturization. Apart from this, periodic self-stabilizing "audits" that "push" the system to a correct state can also help to catch implementation and protocol bugs and, thus, lead to an improvement in the stability of applications.

In this chapter, self-stabilizing content-based routing algorithms for publish/subscribe systems are presented. Self-stabilization is an interesting supplement or even alternative to other fault handling approaches because it provides guarantees to reach a correct state if no fault happens for a long enough time period—regardless which transient faults happened previously. This implies that a self-stabilizing system that has been started in an arbitrary state will eventually reach a correct state in case no fault happens for a sufficiently long time period. With self-stabilizing content-based routing, the configuration of the brokers' routing tables will eventually converge to a legal state such that the publish/subscribe system is able to work correctly again.

The first part of this chapter starts with a definition of the fault model and the notion of a correct publish/subscribe system used (Section 3.2). In the following section, the basic idea and algorithms which rely on a leasing concept designed for particular routing algorithms are presented (Section 3.3). The concrete implementations of self-stabilizing versions of simple, covering-based, identity-based, and merging-based routing are presented together with a discussion of the stabilization time of the respective algorithms. Subsequently, the proposed algorithms are evaluated in a simulation study with respect to message overhead. Finally, generalizations and extensions are presented in Section 3.4, which extend the applicability in order to render arbitrary correct publish/subscribe routing algorithms self-stabilizing. Moreover, advertisements are incorporated and it is shown how the generalization can be applied to peer-to-peer routing.

In the second part of this chapter, a formal analysis of the message complexity in publish/subscribe systems is presented which is then used to determine the overhead induced by self-stabilization (Section 3.5). The analysis opens up the possibility to formally reason about the performance of self-stabilizing content-based routing. Moreover, it represents the first comprehensive theoretical analysis of message overhead in publish/subscribe systems and, thus, provides an alternative to extensive simulations for the evaluation of publish/subscribe systems. The analysis first concentrates on the setting discussed in the experimental evaluation in the first part and presents a formalism based on Markov chains. Then, a generalization is presented which drops several of the restricting assumptions and further broadens the applicability of the analysis.

This chapter closes with a discussion of related work and general issues in Section 3.6 and Section 3.7.

3.2 Foundations

This section starts with defining the notion of a correct publish/subscribe system which is then applied in the definition of a self-stabilizing publish/subscribe system. It closes with a description of the fault model and the introduction of the notion of self-stabilizing content-based routing which will be used in the following.

3.2.1 Correct Publish/Subscribe Systems

The notion of a *correct* publish/subscribe system used in this book builds upon previous work by Mühl, Fiege, and Gärtner who presented a formalization of publish/subscribe systems as a requirement specification [63, 115, 116]. The authors specify publish/subscribe systems using *safety and liveness properties* [100]. Their specification relies on trace-based semantics which are formalized using temporal logic [130]. In the following, an intuitive as well as a formal definition of these properties is given. The formal definitions are formulated with linear temporal logic using the operators "$\square$" (*always*), "$\lozenge$" (*eventually*), "$\bigcirc$" (*next*) and "$\neg$" (*not*) [130].

Definition 5 (Correct Publish/Subscribe System). *A correct publish/subscribe system is a system satisfying the following requirements:*

1. *Safety Properties*

   a) *A notification is only delivered to a client C at most once.*
      $\square[notify(C, n) \Rightarrow \bigcirc\square\neg notify(C, n)]$

b) *A client C only receives notifications which have previously been published.*

$$\Box[notify(C, n) \Rightarrow \exists C' : n \in P_{C'}]$$

c) *A client C only receives notifications it is subscribed for.*

$$\Box[notify(C, n) \Rightarrow n \in N(S_C)]$$

2. *Liveness Property: A client C which is subscribed for a filter f (and does not issue an unsubscription for this filter) will, from some time on, receive every notification n that is published thereafter and is matched by f.*

$$\Box[\Box(f \in S_C) \Rightarrow \Diamond\Box(pub(C', n) \land n \in N(f) \Rightarrow \Diamond notify(C, n))]$$

The definition of a correct publish/subscribe system is useful for the definition of a self-stabilizing publish/subscribe system, since the latter differs from the first only with respect to the safety property: this may temporarily be violated due to faults.

Definition 6 (Self-Stabilizing Publish/Subscribe System). *A publish/subscribe system is self-stabilizing if it satisfies the following requirements:*

1. Eventual Safety Property: *Starting from an arbitrary state, it eventually satisfies (in a bounded time period t) all safety properties given in Definition 5. This is also true for the time after the system has stabilized if no fault occurs.*

$$\Diamond_t S \land \Box[S \Rightarrow \Box S]$$

2. Liveness Property: *The system always satisfies the liveness properties defined in Definition 5.*

It is important to note that it is required that the system, once it reached a legal state, will not reach an illegal state thereafter if no fault occurs. This goes in line with the initial definition of self-stabilization given by Dijkstra [56]. Doing this, systems are not explicitly excluded that are *pseudo-stabilizing* as defined by Burns et al. [28]. Here, a system may switch from one legitimate state to another which may not be recognized by an external observer. In this example, a publish/subscribe system may switch from a correct advanced routing configuration to flooding, where brokers deliver notifications only to their clients if they subscribed for a matching filter. In this case, clients do not observe a difference and the system works correctly since its specification only relies on the interface. However, from the system point of view, this is not desirable since complex routing algorithms are applied to reduce the message complexity. By switching to flooding, the system still remains correct from the perspective of the clients, but the advantage of possibly lower message overhead due to the application of advanced routing algorithms is lost. In general, only a correct routing configuration with respect to the routing algorithm can *guarantee* a correct behavior.

3.2.2 Fault Model

Classical theory on self-stabilization assumes that a system may reach *any* possible *state* due to a transient fault. On the routing layer of a publish/subscribe system, the notion of state comprises the routing tables' entries which are used by the routing algorithm to forward a notification to subscribers as well as all variables stored in RAM which are used by the routing algorithms to draw routing decisions. The state may be corrupted due to transient hardware errors, temporary network link failures resulting in message duplication, loss, corruption, or insertion, arbitrary sequences of process crashes with subsequent recoveries, and arbitrary perturbations of data structures of any fraction of the processes. It is, however, usually assumed that the program code of the routing algorithm itself and data stored in ROM cannot be corrupted.

The routing tables determine the current *routing configuration* of a publish/subscribe system. A *routing algorithm* starts from an eligible *initial routing configuration* and subsequently adapts it. To achieve this *control messages* are propagated through the broker network when clients issue new or cancel existing subscriptions. Intuitively, a routing algorithm is *valid* if it adapts the routing configuration such that the resulting system satisfies the safety and the liveness property of Definition 5. Several content-based routing algorithms are known including simple, identity-based, covering-based, and merging-based routing [115]. These algorithms exist in a peer-to-peer and in a hierarchical variant [33].

In the following, an asynchronous communication model is assumed. Therefore, it is considered a fault if the maximum delay of a communication link exceeds a given limit.

3.2.3 Further Assumptions

In fault model used in this book, the routing configuration can be corrupted arbitrarily by transient faults. Therefore, the routing algorithm must ensure that corrupted routing entries are corrected or deleted from the routing table and that missing routing entries are inserted into the routing table.

Without loss of generality, it is assumed in this chapter that each broker stores the information about its neighbors in its ROM. This ensures that this neighbor information cannot be corrupted. If it would be stored in RAM or on hard disk, it could also be corrupted by a fault. In this case, self-stabilizing content-based routing would need to be layered on top of a self-stabilizing spanning tree algorithm to achieve a self-stabilizing publish/subscribe system. Layered composition of self-stabilizing algorithms is a standard technique which is easy to realize when the individual layers have no cyclic state dependencies [57]. In this case, the stabilization time would be bounded by the sum of the stabilization times of the individual layers.

3.3 Algorithms

In this section, the basic idea of leasing is explained which is used to implement self-stabilizing content-based routing. Then, a description of selected routing algorithms and their self-stabilizing implementations is given. Flooding, simple routing, and the advanced routing algorithms are distinguished which have to be treated separately. Finally, an experimental evaluation of the message overhead due to self-stabilization is presented.

3.3.1 Basic Idea

The basic idea for making content-based routing self-stabilizing is that routing entries are *leased*. A routing entry must be renewed before the *leasing period* π has expired to protect it from being discarded. If a routing entry is not renewed in time, it is removed from the routing table. Interestingly, this approach does not only allow the content-based routing configuration of a publish/subscribe system to recover from internal faults but also from certain external faults. For example, if a client crashes, its subscriptions are automatically removed after their leases have expired. It is required that a subscription has a unique ID which can be generated easily using a unique identifier of the broker together with a timestamp.

To support leasing of routing table entries, a *second chance algorithm* is used. Routing entries are extended by a *flag* which can only take the two values 1 and 0. Before a routing entry is (re)inserted into the routing table, all existing routing entries whose filter has the same ID (as the ID of the filter of the routing entry to be inserted) are removed from the routing table. This is necessary since the IDs and the contents of the routing entries can be corrupted, too. It is assumed that the clock of a broker can only take values between 0 and $\pi - 1$ to ensure that if the clock is corrupted, it can diverge from the correct clock value by at most π. When its clock overruns, a broker deletes all routing entries whose flag has the value 0 from the routing table and sets the flag of all remaining routing entries to 0 thereafter. New subscriptions have the flag set to 1 initially. Hence, it must be ensured that an entry is renewed at least once in π to prevent its expiry. On the other hand, it is guaranteed that an entry which is not renewed will be removed from the routing table after at most 2π.

The renewal of routing entries originates at the clients. To maintain its subscriptions without interruption, a client must renew the lease for each of its subscriptions by "resubscribing" to the respective filter once in a *refresh period* ρ. Resubscribing to a filter is done in the same way as subscribing. In general, π must be chosen to be greater than ρ due to varying link delays. The *link delay* δ is the amount of time needed to forward a message over a communication link and to process this message at the receiving broker. In the model used here, it is considered a fault when δ is not in the range between $\delta_{\min}$ and $\delta_{\max}$. It is important to note that assuming an upper bound for the link delay is a basic assumption of self-stabilization to be able to deal with faults. This is due to the fact that if no bound exists, it is impossible to differentiate between transient and permanent faults. Please note that unsubscriptions are handled regardless of the flags of the routing entries and are, thus, applied to all entries for the given filter ID.

Using this leasing mechanism, an upper bound Δ_u can be given for the stabilization time for idempotent routing algorithms, where resubscriptions have no effect on the routing algorithm:

$$\Delta_u = 2\pi + \rho + d \cdot \delta_{\max} \qquad (3.1)$$

The value of Δ_u is determined by the maximum time, an erroneous routing entry stays in a routing table (2π), plus the maximum time period it then needs until the last subscription is refreshed (ρ), plus the maximum time which is needed to disseminate the last refreshed subscription to all brokers in the system ($d \cdot \delta_{\max}$). After this time period, the routing tables of all brokers are guaranteed to be repaired, such that the publish/subscribe system is back in a correct state. Let d be the *network diameter*, i.e., the length of the longest path a message can take in the broker network.

3.3.2 Routing Algorithms

In this section, self-stabilizing extensions for different content-based routing algorithms are presented. It starts with flooding, which is self-stabilizing by nature, and presents a solution for simple routing and advanced routing algorithms thereafter. More details on the routing algorithms described can be found in [116].

Flooding

The naive implementation of self-stabilizing routing in publish/subscribe systems is *flooding*. With flooding, a broker that receives a notification from a local client forwards it to all neighbor brokers. If it receives a notification from a neighbor broker, it forwards it to all other neighbor brokers. Additionally, it delivers each processed notification to all local clients which subscribed for a matching filter. Flooding only requires a broker to keep state about the subscriptions of its local clients. Therefore, errors in this state can be corrected locally by forcing clients to renew their subscriptions once in a leasing period. This means that $\rho = \pi$ suffices (as stated in Chapter 2, it is assumed that client-broker communication is negligible with respect to time). The main advantage of this scheme is that coordination among neighboring brokers is not necessary. Hence, no additional network traffic is generated. Additionally, new subscriptions become active immediately. Thus, a corrupted or erroneously inserted subscription survives no longer than 2π in a routing table and a missing subscription is reinserted after at most ρ. An erroneously inserted or corrupted notification disappears from the network after at most $d \cdot \delta_{\max}$. Hence, for flooding, the *stabilization time* Δ_f, i.e., the maximum time it takes for the system to reach a legitimate state starting from an arbitrary state, is given by

$$\Delta_f = \max\{2\pi, d \cdot \delta_{\max}\} \tag{3.2}$$

The drawback of flooding is the high message complexity which is independent of the subscribers: filtering is only performed at the local broker of a client which means that notifications may be sent to brokers which do not have any local clients that subscribed for matching filters and which are not needed in order to route these notifications to brokers with subscribed clients.

Simple Routing

Flooding of notifications can be extended to *simple routing* which treats each subscription independently of other subscriptions. A subscription is inserted into the routing table and flooded into the broker network. If a broker B receives a subscription from a local client, it forwards it to all neighbor brokers. If B received it from a neighbor broker, it forwards it to all neighbor brokers except for the sender. Unsubscriptions are handled similarly to subscriptions: if B receives an unsubscription, it removes the respective routing entry. Subsequently, B forwards the unsubscription like a subscription. Thus, simple routing is idempotent to resubscriptions and a subscription is redistributed (flooded) through the broker network when it is renewed by the client. Please note that subscriptions become active only gradually with this routing algorithm, because notifications are forwarded according to routing entries which are installed by subscriptions which need time for dissemination.

Since subscriptions become gradually active, it is a critical issue that the timing assumptions must allow the clients to renew their leases *everywhere* in the network before they expire. How large must π be with respect to ρ in this case? To answer this question, consider two brokers B and B' connected by the longest path a message can take in the broker network. This situation is illustrated in Figure 3.1. Assume that a local client C of B leases a routing table entry of B at time t_0 and renews this lease at time $t_1 = t_0 + \rho$. C's lease causes other leases to be granted all along the path to broker B'. Considering the best and worst cases of the link delay, the first lease reaches B' at time $a_0 = t_0 + d \cdot \delta_{\min}$ in the best case and the lease renewal reaches B' at time $a_1 = t_1 + d \cdot \delta_{\max}$ in the worst case. If C refreshes its leases after time ρ and if network delays are unfavorable, two lease renewals will arrive at B' within at most $a_1 - a_0$. Hence, $\pi > a_1 - a_0$ must hold to ensure that the entry is renewed in time. Thus, the following result for the leasing period is obtained:

$$\pi > \rho + d \cdot (\delta_{\max} - \delta_{\min}) \qquad (3.3)$$

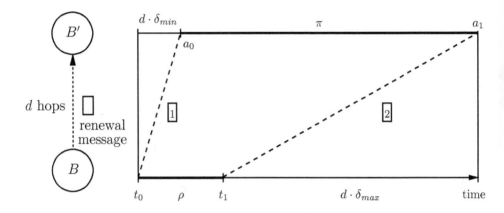

Figure 3.1: Deriving the minimum leasing time

The stabilization time Δ_s depends on the value of π because a corrupted routing entry will stay in a routing table for at most 2π. However, since corrupted or erroneously inserted messages can contaminate the network, a delay of $d \cdot \delta_{\max}$ must be assumed before their processing is finished. Overall, the stabilization time sums up to

$$\Delta_s = d \cdot \delta_{\max} + 2\pi \qquad (3.4)$$

There is a trade off between π and ρ. To obtain a low message overhead ρ should be as large as possible. However, this implies a large value of π, but π should be as small as possible to facilitate fast recovery. In particular, the value of ρ should be chosen such that the additional network traffic that is introduced by ensuring self-stabilization does not degrade the quality of service in an unacceptable way.

Advanced Routing Algorithms

The situation is more complicated if advanced content-based routing algorithms such as identity-based, covering-based, or merging-based routing are applied. In contrast to flooding and simple routing these algorithms are—at least the versions presented so far—not idempotent with respect to resubscriptions. Advanced routing algorithms exploit commonalities between filters to reduce unnecessary forwarding of a new subscription s by a broker B to a neighbor broker B' in case notifications matching s are already forwarded by B' to B due to another subscription matching the same set or a superset of s. Thus, resubscriptions are not forwarded and can not refresh existing routing entries. In the following it is shown, how advanced routing algorithms can be made idempotent with some minor changes. Doing this, it is possible to render them self-stabilizing with the leasing approach already presented.

The term *destination* is used for a broker according to the routing framework proposed in [116]. A destination from the perspective of broker B is a neighbor broker or a local client. The set of B's neighbor brokers is given by N_B. The set of local clients of B is named L_B.

Identity-Based Routing. When a broker B processes a new or canceled subscription s with filter f from destination D, it counts the number i of destinations $D' \neq D$ for which a routing entry with a filter matching the same set of notifications exists in the routing table T_B of B. Depending on the value of i, s is forwarded differently. This way a subscription s for a filter f will not be forwarded to a neighbor broker B' if another subscription with the same filter has been forwarded to B' previously and has not been recalled yet. Details are given in pseudocode notation in Algorithm 1.

This scheme is not idempotent to resubscriptions. If, for example, $i \geq 2$ and one of the identical subscriptions is renewed at B it will not be forwarded to any neighbor broker. If no measures are taken in this case, refreshing subscriptions will, thus, only be forwarded when the identical routing entries have been removed from the routing table. This, however, may lead to missed notifications and, thus, to an incorrect system behavior.

It is proposed that B takes only those subscriptions into account whose flag is 1 when calculating i to make the algorithm idempotent. This simple extension enables to use the leasing mechanism with identity-based routing and gain a stabi-

---

Algorithm 1 Identity-based routing

Broker B received a subscription with filter f from destination D. The variable i holds the number of destinations distinct from D for which a routing entry with an identical filter exists.

1 **if** $(i = 0)$ **then**
2   **if** $(D \in L_B)$ **then**
3   Forward f to all Destinations in N_B
4   **else** $// D \in N_B$
5   Forward f to all Destinations in $N_B \setminus D$
6   **endif**
7 **elseif** $(i = 1 \wedge D' \in N_B)$ **then**
8   Forward f only to D'
9 **endif** $// $ *elseif ($i \geq 2 \vee D' \in L_B$): do not forward f at all*

---

lization time that is equal to that of simple routing. Assume that, due to a fault, an erroneous subscription is fed into the system. It will then take at most $d \cdot \delta_{\max}$ until it has reached the last broker, where it generates a routing entry with its flag set to 1. This routing entry will then persist for at most 2π after which it will be removed by the second chance algorithm. Hence, after time $d \cdot \delta_{\max} + 2\pi$, this fault is corrected.

Another fault that may lead to corruption of a routing table is an erroneously inserted routing entry or a flag that has been toggled. This routing entry may cause unnecessarily forwarded notifications which will stop after at most 2π. More serious is the case, where it blocks "real" subscriptions from being forwarded for at most π (after which its flag is set to 0). Then, after at most ρ, the ordinary refresh mechanism can continue which reinstalls routing entries that may be removed meanwhile due to missing refresh messages. It will take at most $d \cdot \delta_{\max}$ to reach every broker in the system. Thus, from the point on, where the first broker removes a routing entry that has not been refreshed in time, it takes at most $\rho + d \cdot \delta_{\max}$ until the last broker has corrected its routing table.

Since $\pi > \rho$ (Equation 3.3) it is guaranteed that it takes no longer than Δ_i until the system is stable again:

$$\Delta_i = d \cdot \delta_{\max} + 2\pi \tag{3.5}$$

Please note that the way how subscriptions are forwarded depends on the order in which subscriptions with identical filters are renewed after the broker has run the second chance algorithm. However, this order has no impact on the actual forwarding of notifications in the refresh process.

Covering-Based Routing. With covering-based routing, a broker B which receives a subscription from neighbor D with filter f is forwarded according to the value of i as with identity-based routing. However, i is not only calculated by taking identical filters into account but also by checking for filters which match a superset of notifications that f matches. Accordingly, Algorithm 1 can be used, except for that i is calculated using the number of destinations distinct from the sending broker for which a routing entry with an identical or a covering filter exists. Obviously, this routing scheme is not idempotent, too. In the following, two idempotent variants are presented which differ in stabilization time and message complexity. Both use two different i's: one for routing entries with filters that are identical to the filter f of the subscription received (i_i) and another one for routing entries regarding filters that cover f (i_c). A subscription is then forwarded as with identity-based routing, where $i = i_i + i_c$ is used.

In the first variant, only those routing entries are considered when calculating i_i and i_c that have their flag set to 1. This way, a shorter stabilization time compared to the second variant is achieved on the cost of a possibly greater message complexity. This is due to the fact that a covered subscription can be refreshed before the covering subscription is refreshed after a broker executed the second chance algorithm as discussed above. The stabilization time Δ_c in this case is calculated analogous to that of identity-based routing:

$$\Delta_c = d \cdot \delta_{\max} + 2\pi \tag{3.6}$$

The second variant considers all routing entries when calculating i_c regardless their flags. Thus, also routing entries with their flag set to 0 are considered when looking for entries with covering filters (i.e., filters that match a real superset of notifications f covers). Forwarding of subscriptions is accomplished as in the first variant.

By considering also covering filters in routing entries that are flagged with 0 when calculating i_c, this variant avoids unnecessary forwarding of covered subscriptions that may be refreshed later anyway. The case where a covering subscription is refreshed after the covered one, thus, does not lead to unnecessarily forwarded subscriptions of the covered subscription.

This lower message complexity is bought on the cost of a longer stabilization time. Consider the following case, where a subscription with a filter f_a is added due to a fault which covers all other filters and the routing tables of all brokers are initialized due to the same fault short before. In this case, it takes at most time 2π until the routing entries containing f_a are removed (a). Another time ρ is needed at most until the last client has sent a refreshing subscription (b) which needs at most $d \cdot \delta_{\max}$ until it has reached the last broker in the system (c). This results in the stabilization time $\Delta_{c'}$ for the second variant as follows:

$$\Delta_{c'} = \underbrace{2\pi}_{(a)} + \underbrace{\rho}_{(b)} + \underbrace{d \cdot \delta_{\max}}_{(c)} \tag{3.7}$$

In this particular example, another disadvantage of the second variant becomes apparent. While in the first variant no notification would get lost in face of the fault described, notification loss is possible with the second variant until all routing tables are rebuilt.

Merging-Based Routing. Merging-based routing comprises a class of routing algorithms not only one concrete instantiation. Its aim is to further reduce the routing table sizes and the message complexity of subscription forwarding similar to covering-based routing. To achieve this goal, it tries to merge different filters to a new one which replaces the merged filters in the routing tables and is forwarded instead.

In order to make *merging-based routing* self-stabilizing with the approach presented, the refreshing of merged filters must additionally be ensured. As with covering-based routing there is a trade-off between stabilization time and message complexity. Again, two variants of one algorithm are presented.

The focus is on one concrete merging-based routing algorithm implementation which is presented in [116] to exploit its specific mechanisms. This algorithm tries to create a filter f_M that is a *merger* of a set of filters $\{f_1, \ldots, f_n\}$ for the same destination. This merger has to be *perfect*, i.e., $N(f_M) = N(f_1) \cup \ldots \cup N(f_n)$. If it is possible to create a perfect merger, it is forwarded instead of the individual subscriptions. This way it is possible to reduce the routing table size up to 1.

Processing of subscriptions and unsubscriptions is handled similarly to covering-based routing so that analogous mechanisms can be applied. However, it has to be ensured that merged filters are renewed correctly. The routing algorithm stores for every merged filter f_M the set $c(f_M)$ of *constituting filters*. Thus, the second chance algorithm has to be adapted as with covering-based routing above.

Each broker sets the flag of all routing entries containing regular and merged filters and also the flags of the constituting filters to 0 when its clock overruns. At this point, the approach of creating only perfect mergers of the concrete merging-based routing algorithm to lower the message complexity is exploited.

In the first variant, a broker forwards a subscription with a filter that is a constituent filter of a merger in case this merged filter has its flag set to 0. The merged filter is renewed if the last constituent filter has set its flag to 1. Thus, an erroneously inserted merged filter will take time 2π until it is removed from the routing table after it reached the last broker in the system which takes at most $d \cdot \delta_{\max}$. For routing entries with regular filters, the rules from covering-based routing can be applied. The resulting stabilization time Δ_m is equal to that of the first variant of covering-based routing (Equation 3.6):

$$\Delta_m = d \cdot \delta_{\max} + 2\pi \qquad (3.8)$$

To achieve Δ_m, unnecessary forwarding of subscriptions is risked that will later be merged to a new filter which is forwarded instead. This can lead to an increased message complexity. This drawback will be tackled in the second variant.

The second variant is similar to the second variant of the idempotent covering-based routing algorithm. A subscription will never be forwarded if there is a merged filter for which this subscription has a constituent filter. The merged filter itself is refreshed as in the first variant when the last constituent filter has been refreshed. The rationale behind this approach is the assumption that the merger in the routing table is not inserted due to a fault and will be refreshed when the last constituting filter has been refreshed. The main advantage compared to the first variant is that constituent filters are not renewed as regular ones if the merged filter's flag has the value 0. However, this leads to a longer stabilization time in case the merger is the result of a fault. Consider the case, where a merger is erroneously inserted into the routing table of a broker B, claiming that it is constituted by all other filters in the routing table. Moreover, assume that the routing tables of all other brokers have been initialized. Then, refreshing subscriptions will not be forwarded by B until the false merger is removed from its routing table which takes at most time 2π (a). It takes at most another time ρ (b) until the last client has refreshed its subscription afterwards which will take at most time $d \cdot \delta_{\max}$ (c) until it has reached the last broker in the system. This results in a stabilization time $\Delta_{m'}$ of the second variant which is equal to that of the second variant of self-stabilizing content-based routing (Equation 3.7):

$$\Delta_{m'} = \underbrace{2\pi}_{(a)} + \underbrace{\rho}_{(b)} + \underbrace{d \cdot \delta_{\max}}_{(c)} \tag{3.9}$$

Discussion

The concept of leasing is a common way to keep *soft states* which can be used to create flexible and survivable applications [45]. This technique is used in many protocols and algorithms such as the ROUTING INFORMATION PROTOCOL (*RIP*) [108], PROTOCOL INDEPENDENT MULTICAST (*PIM*) [62], and DIRECTED

DIFFUSION [78].

The values of π and ρ depend on the delay of the links in the network. So far, it has been assumed that these values are fixed and equal for every broker in the system. In many scenarios such as large-scale networks or e-home environments, link delays vary a lot among distinct links. Thus, it could be beneficial to incorporate this heterogeneity into the algorithm to gain better stabilization times in parts of the network, where the links delays are lower, such as in local area networks. One approach could be to manually localize the values for the link delays and, thus, all other values that are derived from them. This approach is rather brittle since a network may dynamically become slower or even faster by the time (similar issues arise with mobile brokers). It would then be necessary to calculate the values of π and ρ individually for every subscription, depending on where the subscribers reside. Additionally, π and ρ have to be refreshed the same way as described previously for subscriptions. Taking this approach, the algorithm running on the broker can take advantage of faster links and stabilizes subtrees of the broker topology faster in case the link delays allow for this.

In [141], the authors propose an adaptive approach to refreshing soft state. Their aim is to limit the bandwidth used for control messages by adapting the refresh rate dynamically: if the state a client has to refresh grows, it dynamically reduces the refresh rate to not break this limit. The state-holder has to adapt to the changed refresh rate. Therefore, the authors propose two algorithms: one based on rounds (where clients refresh their state in a round-robin fashion) and one based on prediction. For deterministic self-stabilization it is important to provide strict guarantees in form of a stabilization time that is assured. The adaptive approach does not satisfy this requirement since it adapts the refresh rate to the available bandwidth. It would be an option to define an upper bound for the refresh period and allow for shorter refresh periods if possible. In consequence, parts of the network (e.g., fast local area networks or high-speed backbones) could benefit in form of lower stabilization times. In this case, individual timers for each subscription are necessary. This way it is guaranteed that the maximum refresh period will never be exceeded. If higher refresh rates are possible, the adaptive mechanism would decrease the refresh period again, also if it had been set to a greater value due to a fault.

3.3.3 Evaluation

Introducing soft state in a computer system can increase its robustness significantly. For publish/subscribe systems it additionally fosters the loosely coupled nature inherent to this communication paradigm. This comes at a cost since soft state implies an overhead in control messages which are needed to keep the state in the system.

A discrete event simulation has been carried out to compare self-stabilizing content-based routing to flooding with respect to message complexity. Since the essential issue in self-stabilizing systems is not the average but the *guaranteed* worst-case stabilization time, any experiments on the average stabilization time have been abandoned since this strongly depends on the type and number of faults and the topology of the broker network. Preceding to the discussion of the results, the setup of the experiments is described in the following.

Setup

A broker hierarchy being a completely filled 3-ary tree with 5 levels is considered. Hence, the hierarchy consists of 121 brokers of which 81 are leaf brokers. Since tree is used for routing, this implies a total number of 120 communication links. Similar results to hierarchical routing can be obtained for peer-to-peer routing, too. With hierarchical routing, subscriptions are only propagated from the broker to which the subscribing client is connected towards the root broker. This suffices because every notification is routed through the root broker. Hence, control messages in this scenario travel over at most 4 links. Identity-based routing is used and 1000 different filter classes are considered (e.g., stock quotes) for which clients can subscribe. These classes are exclusive, i.e., an event that matches one class does not match another one.

Subscribers only attach to leaf brokers. Results for scenarios, where clients can attach to every broker in the hierarchy, can be derived similarly. Instead of dealing with clients directly, independent arrivals of new subscriptions with exponentially distributed interarrival times and an expected time of λ^{-1} between consecutive arrivals are assumed. When a new subscription arrives, it is assigned randomly to one of the leaf brokers and one of the available filter classes is randomly chosen. The lifetime of individual subscriptions is exponentially distributed with an expected lifetime of μ^{-1} and each notification is published at a randomly chosen leaf broker. Hence, notifications travel over at most 8 links until they reach a subscriber. The corresponding filter class is also chosen randomly. The interarrival times between consecutive publications are exponentially distributed with an expected delay of ω^{-1}. A constant delay in the broker network of $\delta = 25$ ms including the communication and the processing delay caused by the receiving broker is assumed.

To illustrate the effects of changing the parameters, two possible values are considered for some of the system parameters: for each of the 1000 filter classes, a publication is expected every 1 s (10 s), i.e., $\omega_1 = 1000$ s^{-1} ($\omega_2 = 100$ s^{-1}). The expected subscription lifetime is 600 s (60 s), i.e., $\mu_1 = (600$ s$)^{-1}$ ($\mu_2 = (60$ s$)^{-1}$). Each client refreshes its subscriptions once in 60 s (600 s), i.e., a refresh period of $\rho_1 = 60$ s ($\rho_2 = 600$ s). Since $d = 8$ in the simulation scenario, a leasing period of $\pi_1 = 60.2$ s ($\pi_2 = 600.2$ s) for ρ_1 (ρ_2) is chosen. Hence, a subscription will on average be refreshed 10 (1) times before it is canceled by the subscribing client if $\mu = (600$ s$)^{-1}$. The resulting stabilization time is $\Delta_1 = 120.6$ s ($\Delta_2 = 1200.6$ s). Table 3.1 gives an overview of the system parameters.

The aim is to find out how the system behaves in equilibrium for different numbers of active subscriptions $\mathcal{N}_s$. In equilibrium, $d\mathcal{N}_s/dt = 0$ holds, where $d\mathcal{N}_s/dt = \lambda - \mu \cdot \mathcal{N}_s(t)$, implying $\mathcal{N}_s = \lambda/\mu$. Thus, if $\mathcal{N}_s$ and μ is given, λ can be determined. If the system was started with no active subscriptions, it would be necessary to wait until the system approximately reached the equilibrium before

Parameter	$i = 1$	$i = 2$
ω_i	1000 s^{-1}	100 s^{-1}
μ_i	$(600 \text{ s})^{-1}$	$(60 \text{ s})^{-1}$
ρ_i	60 s	600 s
π_i	60.2 s	600.2 s
$\Rightarrow \Delta_i$	120.6 s	1200.6 s

Table 3.1: System parameter values chosen for evaluation

it is possible to begin the measurements. However, in the example scenario it is possible to start the system right in equilibrium. At time 0, N_s subscriptions are created. For each of these subscriptions, it is determined how long it will live, to which filter class it is assigned to, and at which leaf broker it is allocated. Since an exponential distribution is used for the lifetime, this approach is feasible because the exponential distribution is memoryless. Hence, it does not matter for the remaining lifetime of a subscription how long it is already in the system.

Results

The results of the simulation are depicted in Figure 3.2. A magnification of the most interesting part is shown in Figure 3.3. The results plotted are described in the following:

b_n is the notification bandwidth saved if filtering is applied instead of flooding. The plot depicts b_{s1} and b_{s2} which correspond to the publication rate ω_1 and ω_2, respectively. Because b_s linearly depends on ω, a decrease of ω by a factor of 10 leads to 10 times less saving of notification bandwidth. If there are no subscriptions in the system, $b_{s1} = 116,000 \text{ s}^{-1}$ and $b_{s2} = 11,600 \text{ s}^{-1}$, respectively. These numbers are 4000 s^{-1} and 400 s^{-1} less than the overall number of notifications published per second. This is because with hierarchical routing, a notification is always propagated to the root broker.

b_c depicts the control traffic that is caused by subscribing, refreshing, and un-subscribing clients. It only arises if filtering is used. The figure shows b_{c1}, b_{c2}, b_{c3}, and b_{c4} which result from the different combinations of μ and ρ. The value to which b_c converges for large numbers of subscriptions mainly depends on the refresh period ρ. Thus, b_{c1} and b_{c3} converge to $120,000/\rho_1 = 2000s^{-1}$, while b_{c2} and b_{c4} converge to $120,000/\rho_2 = 200s^{-1}$. The evolution of b_c for numbers of subscriptions in the range between 0 and 200,000 is largely influenced by the value of μ. A small μ such as μ_2 leads to a hump (cf. b_{c3} and b_{c4} in Figure 3.3).

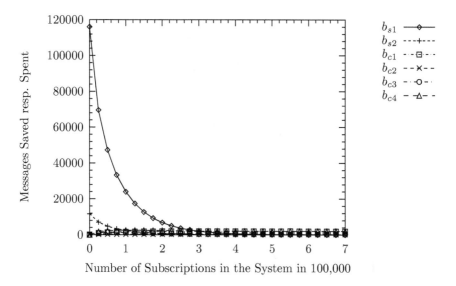

Figure 3.2: Notification bandwidth saved by doing filtering instead of flooding (b_{s1} : $\omega_1 = 1000$ s^{-1}, b_{s2} : $\omega_2 = 100$ s^{-1}) and control traffic caused by filtering and leasing (b_{c1}, b_{c4}, : $\rho_1 = 60$ s, b_{c2}, b_{c3}, : $\rho_2 = 600$ s, b_{c1}, b_{c2} : $\mu_1 = (600$ s$)^{-1}$, b_{c3}, b_{c4} : $\mu_2 = (60$ s$)^{-1}$).

Filtering saves bandwidth compared to flooding if b_s exceeds b_c. The points, where the curves of the respective variants of b_s and b_c intersect are important: if the number of subscriptions is smaller than at the intersection point, filtering is

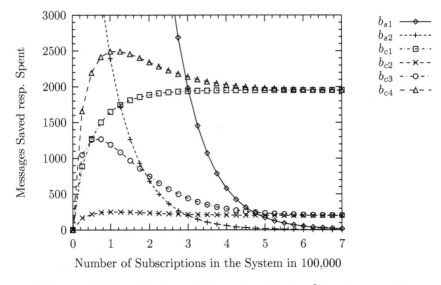

Figure 3.3: Magnification of Figure 3.2 showing $^1/_{40}$ of the y-axis

superior, while for larger numbers flooding is better. For example, the curves of b_{s1} and b_{c1} intersect at about $300,000$ subscriptions. Thus, filtering is superior for less than $300,000$ subscriptions, while flooding is superior for more than $300,000$ subscriptions with the given distributions. Since eight scenarios are considered, there are eight intersection points in Figure 3.3.

Discussion

The setup chosen for the experimental evaluation depicts a worst-case scenario. The assumption that subscribers only connect to leaf brokers increases the message complexity. The same is true for a uniform distribution of subscribers, since publish/subscribe systems are especially beneficial for non-uniform distributions that may, for example, result from locality of subscriptions or differences in the number of clients brokers serve.

3.4 Generalizations and Extensions

In the previous section, several presumptions regarding the routing algorithms and the routing topology have been made. In this section, a generic wrapper algorithm for a very broad range of correct routing algorithms is presented (Section 3.4.1). Due to the general applicability of the wrapper algorithm, it is possible to present a layered approach in Section 3.4.2 in order to integrate advertisements into self-stabilizing content-based routing. The remainder of this section describes how to integrate the generic wrapper algorithm with acyclic as well as with cyclic peer-to-peer routing.

3.4.1 Generic Self-Stabilization Through Periodic Rebuild

In a self-stabilizing system, arbitrary transient faults can occur as pointed out in Section 3.2.2. The only parts that cannot be corrupted are the program code and the data stored in ROM. In general, it is not possible to reason about how a routing algorithm (which works correctly in a fault-free system) behaves when it receives corrupted messages or when it is applied to perturbed routing tables. What can merely be assumed is that it will eventually work correctly again when it is restarted from a legitimate initial routing configuration.

In this section, a generic wrapper algorithm $\mathcal{A}$ for hierarchical routing is presented which renders a publish/subscribe system self-stabilizing, regardless which correct routing algorithm $\mathcal{R}$ it wraps. The only assumptions are that

1. $\mathcal{R}$ has no private state and draws its decision solely on the basis of the respective routing table and the notification or control message it has received.

2. $\mathcal{R}$ terminates after finite time when called[1].

3. Each client refreshes its subscriptions once in a refresh period ρ.

---

[1] Please note that the processing delay caused by the routing algorithm in forwarding a message in the model used is already part of the link delay δ.

The wrapper algorithm periodically rebuilds the routing tables starting from an initial routing configuration that is stored in ROM of each broker. Note that most routing algorithms use an empty initial routing configuration [115]. The algorithm $\mathcal{A}$ can be seen as a periodic precautionary distributed reset [10].

Generic Wrapper Algorithm

Each broker B maintains two routing tables T_B^0 and T_B^1 which are alternately rebuilt on a periodic basis and a flag $a_B \in \{0, 1\}$ which determines which of both routing tables is currently rebuilt. An optimized solution can be implemented with only one table and two flags for every entry indicating to which routing tables the entry belongs[2]. However, notification routing always uses both routing tables to determine the target destinations of a notification. A notification is forwarded to a destination if it matches a routing entry for this destination in any of the two routing tables. If the routing tables are in a correct state, this does no harm.

Since $\mathcal{A}$ wraps $\mathcal{R}$, every call to $\mathcal{R}$ is intercepted by $\mathcal{A}$. $\mathcal{A}$ determines which routing table the next call of $\mathcal{R}$ operates on as described in the following. For every subscription from a local client of B, $T_B^{a_B}$ will be used. If control messages are generated by $\mathcal{R}$ in reaction to the subscription, they will be tagged with a_B. Accordingly, when a broker B' receives a control message tagged with x from a neighbor broker, then $T_{B'}^x$ will be used by $\mathcal{R}$ for this call. If B receives an unsubscription (from any destination), it applies it to both routing tables. This is done to avoid unnecessary forwarding of notifications. The control messages generated by $\mathcal{R}$ for both routing tables in reaction to the unsubscription are then forwarded to the respective destinations, all tagged with a_B.

The periodical rebuild is triggered by a modulo clock every π on the root broker R. The rebuild sets $a_R \leftarrow \neg a_R$. Then, it initializes $T_R^{a_R}$ with the initial routing configuration stored in ROM and propagates a switch(a_R) message to all of its neighbors. Similarly, when a broker B' receives a switch(x) message from a neighbor, it sets $a_{B'} \leftarrow x$, initializes $T_{B'}^{a_{B'}}$ and forwards a switch(x) message to all other

---

[2]Actually, this is the way self-stabilizing identity-based routing has been implemented in Section 3.3.2.

neighbors.

If a subscription (unsubscription) is issued twice by a client between two consec-utive switch messages without an intervening unsubscription (subscription), this could raise a problem because $\mathcal{R}$ might not tolerate resubscriptions. To avoid this potential problem the use of another wrapper algorithm as described later in this section is proposed.

Correctness. Before the correctness of the presented scheme is shown, the prepara-tory Lemma 1 will be proven which determines a lower bound for the value of π. Then, with the help of Theorem 1 it will be shown that, if this bound holds, $\mathcal{A}$ renders content-based routing in publish/subscribe systems self-stabilizing.

Lemma 1. *In a correct system, if $\pi > 2 \cdot d \cdot \delta_{\max}$, no "old" control messages tagged with x can arrive at any broker after the root broker issued the next "new" switch(x) message.*

Proof. Old control messages tagged with x disappear at most $d \cdot \delta_{\max}$ after the last broker has received the switch($\neg x$) message. This means that at most $2 \cdot d \cdot \delta_{\max}$ after the root broker has sent the switch($\neg x$) message no old control messages tagged with x can arrive. Since π is greater than this value, only new control messages tagged with x can arrive at any broker after the next "new" switch(x) message is issued by the root broker (Figure 3.4). □

Theorem 1. *When the wrapper algorithm is applied and $\pi > \rho + 2 \cdot d \cdot \delta_{\max}$ holds, content-based routing is self-stabilizing and the stabilization time Δ_g is given by*

$$\Delta_g = 2\pi + d \cdot \delta_{\max} \tag{3.10}$$

Proof. For the correctness, it is necessary to show that (1) the system stays in a correct state if it is currently in a correct state (*closure*) and that (2) the system will eventually enter a correct state if it is currently in an incorrect state (*convergence*).

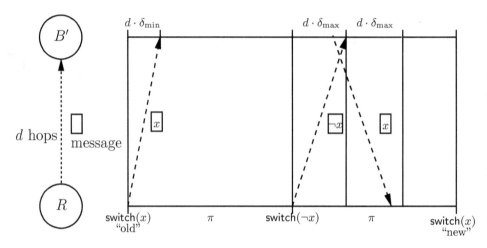

Figure 3.4: Choosing π such that "old" and "new" control messages do not interleave

(1) For the system to stay in a correct state, it is necessary to ensure that (1a) at each broker the rebuild process of the routing table which is currently rebuilt is completed before the next switch message is received, that (1b) the rebuild is based only on new control messages, and that (1c) all new control messages are received after the respective switch message.

(1a) This means that at each broker the time period between two consecutive switch messages must be large enough to ensure that all necessary control messages are received in time. The time difference at which two brokers receive the same switch message cannot be greater than $d \cdot \delta_{\max}$. At all brokers, the clients need at most ρ to reissue all their subscriptions after the broker has received the switch message. The resulting control messages need at most $d \cdot \delta_{\max}$ to travel through the broker network. Therefore, $\pi \geq \rho + 2 \cdot d \cdot \delta_{\max}$ must hold to guarantee that at each broker the rebuild is complete before the next switch message is received.

(1b) By Lemma 1 and the fact that $\pi \geq \rho + 2 \cdot d \cdot \delta_{\max}$.

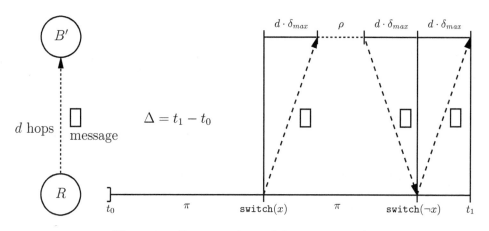

Figure 3.5: Deriving the stabilization time Δ_g

(1c) Due to the FIFO-property of the communication channels and the fact that the topology is acyclic, a broker B' can only receive control messages and (un)subscriptions of local clients tagged with x after B' received the corresponding switch(x) message.

(2) Starting from an arbitrary state, every broker receives the next switch message after at most $\pi + d \cdot \delta_{\max}$. This message causes the receiving broker to reinitialize one of its two routing tables. Due to (1) it is guaranteed that this routing table will be completely rebuilt before the subsequent switch message is received. This second switch message is received by all brokers at most $2\pi + d \cdot \delta_{\max}$ from the beginning. It causes the other routing table to be reinitialized. After all brokers have received and processed the second switch message, the system is guaranteed to be in a correct state again. This is because at all brokers the one routing table is completely rebuilt, while the other is reinitialized.

Therefore, the stabilization time Δ_g is $2\pi + d \cdot \delta_{\max}$ (see Figure 3.5).

$\Box$

Handling Repeated (Un)subscriptions

The self-stabilizing algorithms presented rely on repeated refresh messages for subscriptions. Routing algorithms are not necessarily robust to resubscriptions although the algorithms investigated so far had this property. To be able to support every correct routing algorithm $\mathcal{R}$ it is, thus, possible to use another wrapper algorithm $\mathcal{I}$ which makes $\mathcal{R}$ idempotent by keeping a list of all subscriptions that have been received in the active leasing period. If $\mathcal{I}$ is called with a repeated subscription, it silently discards it and immediately returns from the procedure sub(), according to the specification of the routing framework introduced in [115].

Overhead Considerations

The generic wrapper algorithm is equivalent to the algorithm proposed for identity-based routing and very similar to the first variant of the idempotent algorithms for covering-based and merging-based routing presented in Section 3.3.2. Similarly to the generic wrapper algorithm, they only consider routing table entries with their flag set to 1, simulating two routing tables thereby. As shown previously, the complete rebuild of the routing tables can comprise a higher message complexity since the order in which routing entries are refreshed can have a significant influence on the way (un)subscriptions are forwarded. With covering-based routing, for example, it is always beneficial if the largest filter is refreshed first since covered filters, if at all, only need to be forwarded to fewer destinations subsequently.

3.4.2 Advertisements

Without using advertisements, the notification service has to take care that for every subscription there will eventually be a respective routing entry (depending on the routing algorithm used) at every broker's routing table. This is due to the fact that a publisher may connect to any broker in the system. The motivation of *advertisements* [33] is to reduce control traffic due to (un)subscriptions by actively announcing publishers with their respective set of notifications they may produce.

This is, of course, only effective if the rate at which publishers cease or new publishers arrive is low, i.e., if the subscribers behave more dynamically than the publishers do.

Every publisher "advertises" the set of possible notifications it may publish by publishing advertisement filters. These filters must match all notifications this publisher is going to produce and is disseminated in the broker network. Besides the routing table for notifications, hence, every broker also holds a routing table solely for advertisements. When a broker receives a new advertisement it forwards it according to the routing algorithm used, which can be any of the subscription routing algorithms. Cancellations of advertisements (i.e., unadvertisements) are handled analogous to unsubscriptions.

When advertisements are used, subscriptions only need to be forwarded to destinations for which an overlapping advertisement exists. For every other destination it is unnecessary to forward the subscription since no matching notification will be published by any client in this direction.

In the following, it is shown how to incorporate advertisements into self-stabilizing content-based routing using the generic wrapper algorithm for both routing tables. Other routing algorithms can be used with only minor modifications.

Combining Advertisements with Subscriptions

Publish/subscribe systems that use advertisements contain a dependency between the routing tables built by advertisements (i.e., the *subscription routing tables*) and those built by subscriptions (i.e., the *notification routing tables*) which implies an *order* in which the routing tables have to be rebuilt (Figure 3.6): the notification routing table can only be repaired when the subscription routing table is in a correct state. This is due to the fact that subscriptions are routed according to the subscription routing table like notifications are routed using the notification routing table on each broker.

The goal is to create a self-stabilizing routing stack which consists of two layers: the subscription and the notification routing tables. The latter gets its input from

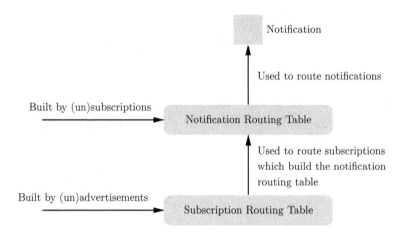

Figure 3.6: Dependency between different routing layers in publish/subscribe systems which use advertisements

the first. According to Dolev [57], it is possible to create this algorithm stack by simply composing self-stabilizing algorithms for both layers, since there are no cyclic dependencies between them. This approach is called *fair composition* and presumes that each algorithm step is executed infinitely often. Doing this, it is not necessary for the algorithm which stabilizes the notification routing table to detect whether the subscription routing table is stable—if the latter is stable, the first will stabilize because it is self-stabilizing. For the stabilization time Δ_a, this means that it equals the sum of the stabilization times of both layers. When using the generic wrapper algorithm on both layers this is

$$\Delta_a = 2 \cdot (2\pi + d \cdot \delta_{\max}) \tag{3.11}$$

3.4.3 Acyclic Peer-to-Peer Routing

The generic wrapper algorithm presented relies on a distinct root broker R which periodically broadcasts a switch(x) message and, thus, synchronizes all brokers.

This mechanism can easily be transferred to acyclic peer-to-peer routing, because R can be an arbitrary dedicated broker in the network. The fact that (un)subscriptions are only forwarded to R has never been exploited when designing the generic wrapper algorithm. Thus, it is compatible with peer-to-peer routing, too.

3.4.4 Cyclic Peer-to-Peer Routing

In the previous sections, only acyclic routing topologies have been considered (i.e., hierarchical and acyclic peer-to-peer routing) with exactly one path between two distinct brokers. Acyclic topologies simplify the task of preventing duplicates. In a fault-free scenario a notification is guaranteed to arrive no more than once at a broker since forwarding cycles are avoided. Carzaniga proposes a routing algorithm for acyclic peer-to-peer topologies which relies on reverse path forwarding [33]. Thereby, the acyclic structure is broken up into a structure, where there is always only *one* shortest path between two brokers, preventing forwarding cycles this way. Therefore, each broker B_x needs a routing table which stores the next broker on the shortest path between B_x and B_y in the system.

The routing algorithms that are used on top of this cyclic routing topology are the same that are used for cyclic topologies. Thus, it is possible to use the generic wrapper algorithm here, too.

3.5 Analysis of Self-Stabilizing Content-Based Routing

The focus in the first part of this chapter laid on algorithms for rendering content-based routing self-stabilizing. In this second part, the first comprehensive analysis of the message complexity of hierarchical publish/subscribe systems including self-stabilization is developed, providing an alternative to extensive simulations. The analysis is based on continuous time birth-death Markov chains and investigates the characteristics of publish/subscribe systems in equilibrium. Closed analytical solutions will be given in a simplified setting for the sizes of routing tables, for the overhead required to keep the routing tables up-to-date, and for the leasing overhead required for self-stabilization. Later, many restricting presumptions are dropped and a generalization is presented which facilitates the analysis of much more complex scenarios.

3.5.1 Assumptions

The common model of a publish/subscribe system is assumed made up by a set of cooperating brokers forming an overlay network. In this model, transient faults can affect the communication channels connecting neighboring brokers as well as the state of the brokers. The focus is on self-stabilizing routing tables on the publish/subscribe layer and it is assumed that the overlay network topology is static and, thus, stored in ROM (i.e., it cannot be corrupted). An alternative would be layering self-stabilizing routing algorithms on top of a self-stabilizing broker overlay topology. This is a standard technique which is easy to realize if both layers have no cyclic state dependencies [57]. However, layering self-stabilizing content-based routing on top of a self-stabilizing broker overlay tree has not been taken into account. Doing this would demand further efforts that would require additional information about the self-stabilizing algorithm which is used on the overlay layer. For example, a fault may completely change the topology which may have an impact on the number of messages sent.

Model

A broker hierarchy is considered that forms a complete m-ary tree with k levels, where $m \geq 1$ and $k \geq 1$. The number of brokers on the i-th level $0 \leq i \leq k-1$

equals m^i. Hence, the *total number of brokers* $\mathcal{N}_B$ in the hierarchy is

$$\mathcal{N}_B = \sum_{i=0}^{k-1} m^i = \frac{m^k - 1}{m - 1} \tag{3.12}$$

and the *number of leaf brokers* $\mathcal{N}_L$ is

$$\mathcal{N}_L = m^{k-1} \tag{3.13}$$

implying the *total number of communication links l* of

$$l = \mathcal{N}_B - 1 \tag{3.14}$$

since a tree is used for routing. A hierarchy for $m = 2$ and $k = 4$ is shown in Figure 3.7.

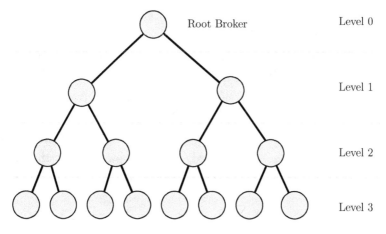

Figure 3.7: An exemplary broker topology

It is initially assumed that subscribers are only attached to leaf brokers to simplify the analysis. Results for scenarios, where clients can be attached to every broker in the hierarchy, can be derived similarly through superposition. Please note that this assumption is a worst-case assumption because the traffic increases if the subscribers are only attached to the leaf brokers. The communication between the subscribers and their local broker is assumed to be local and is, thus, not considered in the analysis.

Instead of dealing with clients directly, independent arrivals of new subscriptions are assumed, where the interarrival times of subscriptions are exponentially distributed with an expected delay of $(\mathcal{N}_L \cdot z \cdot \lambda)^{-1}$ between consecutive arrivals. When a new subscription arrives, it is assigned randomly to one of the $\mathcal{N}_L$ leaf brokers using a uniform distribution, choosing randomly one of z different filter classes using also a uniform distribution. This means that from the viewpoint of a single leaf, the arrival times of subscriptions regarding a specific filter are exponentially distributed with an expected interarrival time of λ^{-1}. The lifetime of individual subscriptions is exponentially distributed with an expected lifetime of μ^{-1}.

For the analysis it is assumed that hierarchical routing is used. The applicability of the analysis with other routing algorithms is discussed in Section 3.5.5. This implies that subscriptions only have to be propagated towards the root broker as every notification is routed through the root broker. First, it is assumed that all notifications are published by a single producer which is connected to the root broker. This assumption is dropped later and the general case is discussed, where the publisher can attach to any broker. Regarding the notifications produced, an average publication rate ω is considered. When a new notification is published, the corresponding filter class is randomly chosen using a uniform distribution.

Example Scenarios

Throughout this analysis, there will be references to exemplary settings in order to illustrate the theoretical results. In these settings, the broker hierarchy is a 3-ary tree with 5 levels ($m = 3$ and $k = 5$). Hence, there are $l = 120$ communication links and $\mathcal{N}_B = 121$ brokers of which $\mathcal{N}_L = 81$ are leaf brokers. Since hierarchical routing is used, this means that control messages (i.e., forwarded (un)subscriptions) have to travel over at most 4 links. It is assumed that there are $z = 1000$ different filter classes for which clients can subscribe.

To illustrate the effect of a variation in the parameters, plots are created for eight different settings if appropriate, where for each filter class one $(1/10)$ publication is expected per second, i.e., $\omega = 1000$ s^{-1} ($\omega = 100$ s^{-1}). For the subscriptions, an expected subscription lifetime of 600 s (60 s), i.e., $\mu = (600 \text{ s})^{-1}$ ($\mu = (60 \text{ s})^{-1}$) is assumed. Furthermore, a refresh period of $\rho = 60$ s ($\rho = 600$ s) is assumed.

The minimum delay for one hop in the overlay network including the communication delay on a link and the processing delay of a broker is $\delta_{\min} = 1$ ms and the maximum delay is $\delta_{\max} = 251$ ms. The diameter d equals $k - 1$ in this setting, $d \cdot \delta_{\min} = 4$ ms and $d \cdot \delta_{\max} = 1.004$ s, which implies a leasing period of $\pi = \rho + d \cdot (\delta_{\max} - \delta_{\min}) = 61$ s ($\pi = 601$ s) for $\rho = 60$ s ($\rho = 600$ s). This means, for example, that a subscription will on average be refreshed 10 (1) times before it is canceled by the subscribing client if $\rho = 60$ s ($\rho = 600$ s) and $\mu = (600 \text{ s})^{-1}$.

These settings resemble the settings chosen for the simulation in Section 3.3.3 and can, hence, be used to verify the results gained there.

3.5.2 Modeling State Distribution Using Markov Chains

For the analysis, a *continuous time birth-death Markov chain* is used that corresponds to an $M/M/\infty$ queuing system which is also known as *responsive server* (Figure 3.8) [95, Chapter 3.4]. For every leaf broker and filter class, an independent Markov chain is considered, where the birth rate $\lambda_i = \lambda$ does not depend on the current state of the chain and where the death rate depends on the state and is given by $\mu_i = i \cdot \mu$. In the exemplary setting, thus, $81,000$ independent Markov chains are considered. A Markov chain is in state i when the respective leaf has i simultaneously active subscriptions for the respective filter class.

Since the equilibrium is of major interest, the focus is on the stationary distribution of the states. It is easy to see that in the scenario assumed the condition for a stationary distribution is met because a k exists, where $\lambda_i/\mu_i < 1 \ \forall i \geq k$. In

Figure 3.8: State transition rate diagram of $M/M/\infty$ queuing system

this scenario, where $\lambda_i = \lambda$, $\mu_i = i \cdot \mu$, and p_i is the probability that a broker has exactly i subscriptions, the following holds according to [94]:

$$p_i = \frac{e^{-\frac{\lambda}{\mu}}}{i!} \left(\frac{\lambda}{\mu} \right)^i \tag{3.15}$$

Let $\mathfrak{n}$ be the *number of subscriptions per leaf and filter class*. In this case, the expected value $\bar{\mathfrak{n}}$ in equilibrium is

$$\bar{\mathfrak{n}} = \frac{\lambda}{\mu} \tag{3.16}$$

Hence, in equilibrium a leaf has on average λ/μ subscriptions for one filter class. Figure 3.9 depicts the state distribution of one of the Markov chains, where it is easy to see how the number of subscriptions per filter class in equilibrium grows with an increasing total number of subscriptions in the system.

Finally, λ/μ can be substituted by $\bar{\mathfrak{n}}$ in Equation 3.15:

$$p_i = \frac{e^{-\bar{\mathfrak{n}}} \cdot \bar{\mathfrak{n}}^i}{i!} \tag{3.17}$$

In the context of this analysis, the probability p_0 that a broker has no subscription plays a central role:

$$p_0 = e^{-\bar{\mathfrak{n}}} \tag{3.18}$$

Since the *expected number of active subscriptions in the system* $\bar{\mathfrak{N}}$ is

$$\bar{\mathfrak{N}} = \bar{\mathfrak{n}} \cdot \mathcal{N}_L \cdot z \tag{3.19}$$

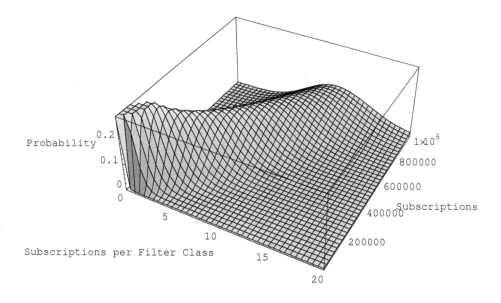

Figure 3.9: Distribution of the states of one Markov chain in equilibrium for $\omega = 1000\ \text{s}^{-1}$, $\rho = 60\ \text{s}$, and $\mu = (600\ \text{s})^{-1}$

It is possible to determine p_i by setting the system parameter $\bar{\mathfrak{N}}$.

3.5.3 Message Complexity

In this section, the message complexity of self-stabilizing flooding and identity-based routing is investigated. To compare the efficiency of both routing algorithms, the number of messages is analyzed which are produced with each of the algorithms and it is calculated, in which settings either algorithm is preferable.

The message complexity of flooding b_f is simply the number of notification messages sent over all links:

$$b_f = \omega \cdot l \tag{3.20}$$

The messages created when using self-stabilizing identity-based routing consist of *notification messages* b_n and *control messages* b_c consisting of refresh messages

67

due to *leasing* b_l and *toggle messages* b_o.

Refresh Messages. A *refresh message* is created every time period ρ for each ⟨subscription, leaf⟩ combination. As already described in Section 3.3.1 the brokers apply a second chance algorithm and refresh messages are only forwarded to parent brokers if the respective subscription has had its flag set to 0 before. This way, a refresh message is created at most once in period ρ for each subscription and forwarded over at most all links up to the root broker. Again, refresh messages sent by the clients to their local broker are not counted as it is assumed that this is free of charge local communication.

Toggle Messages. When a broker receives an unsubscription for the routing entry containing a certain filter class, it "toggles" its state from subscribed to not subscribed for this filter class. The opposite is true when a broker receives a subscription with a certain filter class for which no routing entry yet exists. In both cases, the broker sends a (un)subscribe message to its parent broker (if it is not the root broker) and causes a *toggle message* this way. One such toggle can generate up to $k-1$ subsequent toggle messages, e.g., if the first client subscribes for a filter class to which no other client is subscribed to in the subtree that is rooted in a broker on the first level.

Accordingly, the number of messages b saved when applying identity-based routing instead of flooding is

$$b = b_f - (b_n + b_l + b_o) \tag{3.21}$$

Notification Messages

To analyze the number of notification messages that occur it is necessary to consider the expected number of remote routing entries, since every notification message will be sent over as many links as there are remote routing entries in the system for the matching filter class.

Leaf brokers can only have local routing entries because hierarchical routing is used. Every non-leaf broker has a remote routing entry for each ⟨filter class, child broker⟩ combination if there is a client subscribed to this filter class that is connected to a leaf broker in the tree rooted in the respective child broker. If a broker receives a notification from its parent broker, it forwards it only to those child brokers for which there is a routing entry for a combination consisting of this child broker and a matching filter class. There can be no more than $l \cdot z$ remote routing entries in the system (i.e., $120,000$ in the exemplary setting).

On the i-th level there are m^i brokers and each subtree rooted in one of the m child brokers of a broker on the i-th level contains m^{k-2-i} leaf brokers. The probability that all these leafs have no subscription for a certain filter class is $p_0^{m^{k-2-i}}$. Hence, the expected number of occupied remote routing entries x in the system is

$$x = z \cdot \left(l - \sum_{i=0}^{k-2} m^{i+1} \cdot p_0^{m^{k-2-i}} \right) \tag{3.22}$$

When the number of subscriptions in the system $\mathfrak{N}$ grows, x converges to $l \cdot z$ because p_0 approaches zero in this case (Figure 3.10). This is consistent with the intuition that when the expected number of subscriptions in the system reaches a certain point, it is expected that every leaf broker has a subscription for each filter class, since the subscriptions are uniformly distributed to all leaf brokers.

If filtering is applied, a published notification traverses on average x/z links. Hence, the bandwidth used by notification filtering is given by

$$b_n = \omega \cdot \frac{x}{z} \tag{3.23}$$

The portion of the notification traffic y that is saved by applying filtering instead of flooding is given by

$$y = \frac{b_f - b_n}{b_f} = \frac{l \cdot z - x}{l \cdot z} = 1 - \frac{x}{l \cdot z} \tag{3.24}$$

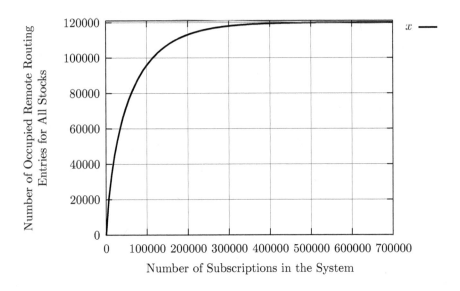

Figure 3.10: Number of remote routing entries

and as expected, y approaches 0 for large $\overline{\mathfrak{N}}$.

The bandwidth used for notification forwarding that is saved by applying filtering instead of flooding b_s is

$$b_s = b_f - b_n = \omega \cdot \left(l - \frac{x}{z}\right) \tag{3.25}$$

Figure 3.11 shows how the plot of b_s approaches zero for the two different publication rates.

Leasing

The bandwidth used for leasing depends on the refresh period ρ as the clients refresh their active subscriptions once in a period. For each remote routing entry that is refreshed, one refresh message is sent. Hence, the bandwidth used for

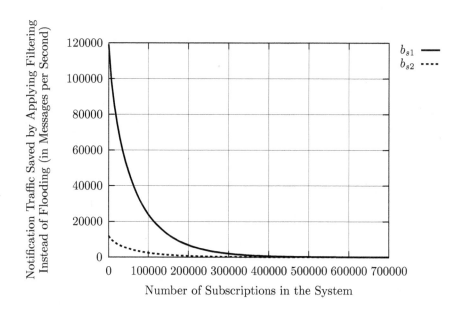

Figure 3.11: Saved notification bandwidth by applying filtering instead of flooding (for $b_{s1} : \omega = 1000$ s^{-1} and $b_{s2} : \omega = 100$ s^{-1})

leasing b_l is

$$b_l = \frac{x}{\rho} \tag{3.26}$$

It is obvious, that the number of routing entries refreshed per second converges to $z \cdot l/\rho$ since x converges to $z \cdot l$ for large n. In the exemplary setting, b_l, thus, approaches 2000 (200) messages per second for $\rho = 600$ s ($\rho = 60$ s) for large numbers of subscriptions (Figure 3.12).

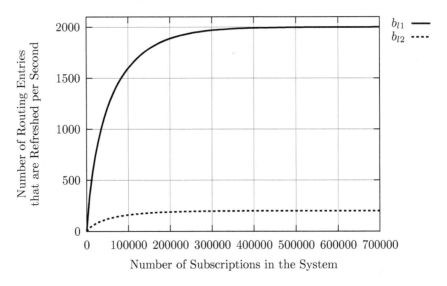

Figure 3.12: Bandwidth used by leasing (for $b_{l1} : \rho = 600$ s and $b_{l2} : \rho = 60$ s)

Toggling

The second type of control message caused by (un)subscriptions to keep the routing tables of the brokers up-to-date are the toggle messages. Here, only two states of the Markov chain have to be considered (Figure 3.13): a leaf broker either has a subscription for a filter class (state 1) or it has not (state 0). In consequence, two state transitions have to be considered: (i) a leaf toggles from state 0 to state 1 because a first subscription for this filter class arrived and (ii) a leaf toggles from state 1 to state 0 because the last subscription for this class disappeared.

Figure 3.13: Markov chain for leaf toggling

Please note that toggle messages that are caused by brokers that change their state from state 1 to state 0 can be disregarded if unsubscriptions are only used to modify local routing entries and are not forwarded accordingly. In this case, removing remote routing table entries at other brokers is left to the second chance algorithm. This would reduce the number of toggle messages on the cost of an increase in the notification traffic. This is due to the fact that it may then happen that notifications are sent to a broker because of remote routing entries that had been inserted due to a subscription which has already been revoked meanwhile by the respective subscriber.

To determine the bandwidth that is used for leaf toggling, it is necessary to find out how often a leaf toggles between the two states and how many messages a toggle creates. First, it is determined how many messages a toggle costs on average.

For the number of tree levels $k = 1$, a toggle message is never sent because all clients are attached locally to a single broker. For $k = 2$, there are m leaf brokers with clients and a root broker to which the leaf brokers are directly connected. In this case, a single toggle message is sent for each toggle. Now, the more interesting case where $k \geq 3$ is discussed.

If a leaf broker toggles in either direction for a filter class, it always sends a message to its parent broker. An inner broker B sends a toggle message that it received from one of its child brokers to its parent broker if all leaf brokers that are located in the subtree rooted in B (except for that leaf which is responsible for the toggle message) are in state 0 for this filter class. If B is on the i-th level, this affects $m^{k-1-i} - 1$ leaf brokers. Hence, the probability that B sends a toggle message to its parent broker is $p_0^{m^{k-1-i}-1}$. Thus, the expected number of control messages per toggle o is

$$o = \sum_{i=1}^{k-1} p_0^{m^{k-1-i}-1} = \sum_{i=0}^{k-2} p_0^{m^i-1} \qquad \forall k \geq 3 \qquad (3.27)$$

Note that o quickly approaches 1 when $\bar{\mathfrak{N}}$ grows (Figure 3.14) which is intuitive, since with a growing number of subscriptions the probability that other brokers toggled decreases.

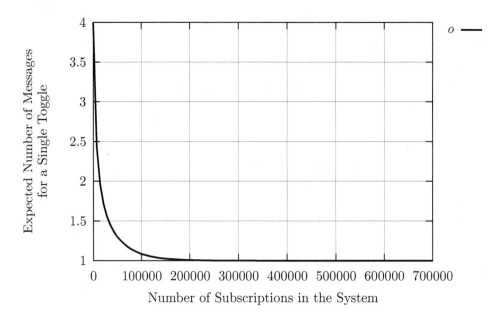

Figure 3.14: Number of messages per toggle

Now, the expected number of toggles is derived. Since the birth rate $\lambda' = \lambda$ and the probabilities $p'_0 = p_0$ and $p'_1 = 1 - p'_0 = 1 - p_0$ that the broker is in state 0 and state 1, respectively, is known, it is possible to derive the death rate μ':

$$\mu' = \lambda' \cdot \frac{p_0'}{p_1'} = \lambda \cdot \frac{p_0}{1 - p_0} \tag{3.28}$$

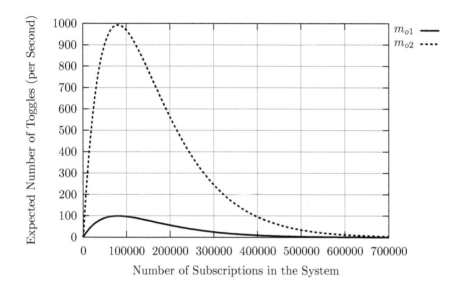

Figure 3.15: Number of toggles per second (for $m_{o1} : \mu = (600 \text{ sec})^{-1}$ and $m_{o2} : \mu = (60 \text{ s})^{-1}$)

If a leaf enters state 0, the expected time it stays in this state before it toggles to state 1 is λ'^{-1}. The reason for this is that the interarrival times are exponentially distributed with mean λ^{-1}. Similarly, if the leaf enters state 1, the expected time it stays in state 1 is μ'^{-1} due to the exponential distribution of the subscription lifetimes. Thus, each leaf will toggle on average two times in time period $\lambda'^{-1} + \mu'^{-1}$ for each of the z filter classes. Thus, the toggle rate $m_o^f(B_l)$ of a leaf B_l for one filter class f is given by

$$
\begin{aligned}
m_o^f(B_l) &= \frac{2}{\lambda'^{-1} + \mu'^{-1}} = \frac{2}{\frac{1}{\lambda'} + \frac{1}{\lambda'} \cdot \frac{1-p_0}{p_0}} = \frac{2 \cdot \lambda'}{\frac{p_0 + 1 - p_0}{p_0}} \\
&= 2 \cdot \lambda' \cdot p_0 = 2 \cdot \bar{n} \cdot \mu \cdot e^{-\bar{n}}
\end{aligned}
\tag{3.29}
$$

The toggle rate, thus, depends on the toggle rate of a leaf broker for one filter class, on the expected number of subscriptions this broker maintains, and the expected lifetime of these subscriptions. For example, the longer the subscriptions last in the system (i.e., the lower the death rate μ), the lower is the toggle rate.

Since there are $\mathcal{N}_L$ leafs in the hierarchy and there are z different filter classes, the following toggle rate m_o is obtained for the system:

$$
m_o = \mathcal{N}_L \cdot z \cdot m_o^f = \mathcal{N}_L \cdot z \cdot 2 \cdot \bar{n} \cdot \mu \cdot e^{-\bar{n}}
\tag{3.30}
$$

In the exemplary setting, m_o has its maximum of approximately 99 (993) toggles per second for $80,998$ subscriptions (Figure 3.15). The expected bandwidth used for toggling b_o is derived by multiplying o with m_o:

$$
b_o = o \cdot m_o
\tag{3.31}
$$

The maximum of 117 (1174) messages per second is reached in the exemplary setting with $57,382$ subscriptions and the expected subscription lifetime $\mu^{-1} = 60$ s ($\mu^{-1} = 600$ s) (Figure 3.16).

Additional Bandwidth for Publishers at Leaf Brokers

In the model as described in Section 3.5.1, it is assumed that all notifications are published at the root broker. This is a best-case assumption because with hierarchical routing every notification published has to be sent at least to the root broker—even if there is no subscriber at all in the system. If this assumption is relaxed and the case is considered, where notifications are instead published at the

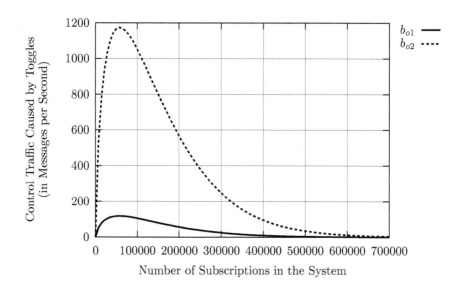

Figure 3.16: Bandwidth used by toggling (for b_{o1} : $\mu = (600 \text{ s})^{-1}$ and b_{o2} : $\mu = (60 \text{ s})^{-1}$)

leaf brokers, one recognizes that a published notification may traverse more links than if it were published at the root, which resembles the worst case. However, this concerns only those $k-1$ links on the path from the publishing leaf to the root broker. A message sent from a broker B on level i to its parent broker on level $i-1$ $(1 \leq i \leq k-1)$ is *additional* if no leaf in the subtree rooted in B has a subscription for the respective filter class. In this case, if the same message would have been published at the root broker, the message would not have traversed this link. The number of leaf brokers in a subtree rooted in a broker on level i equals m^{k-1-i}. Hence, the probability that none of the leafs has a subscription for the respective filter class is $p_0^{m^{k-1-i}}$. The publication rate holds for all leaf brokers and the expected number of additional messages is equal for all of them. Thus, the expected additional bandwidth b_p is given by

$$b_p = \omega \cdot \sum_{i=1}^{k-1} p_0^{m^{k-1-i}} = \omega \cdot \underbrace{\sum_{i=0}^{k-2} p_0^{m^i}}_{=p_0 \cdot o} = \omega \cdot p_0 \cdot o \qquad (3.32)$$

and plotted in Figure 3.17. The relationship between the expected additional number of notifications and the expected toggle messages in this case comes from the fact that a toggle message is generated if all brokers on the same level i are in state 0 for the respective filter class. A notification sent to the root, however, is additional if not only the other brokers on the same level i but also the publishing broker itself is in state 0. It is, thus, necessary to multiply the probabilities accumulated in Equation 3.27 with p_0.

For all values of $\bar{\mathfrak{N}}$, the bandwidth saved when applying filtering instead of flooding, i.e., $b_f - b_n$, is at most $l/(k-1)$ times the additional bandwidth b_p. In the example setting, this means a maximum of 30, i.e., publishers at leaf brokers add no more than 3.3 % of the traffic saved when applying filtering instead of flooding. Thus, the assumption that notifications are published at the root broker does not have a significant influence on the bandwidth needed when applying filtering and can be disregarded for growing $\bar{\mathfrak{N}}$.

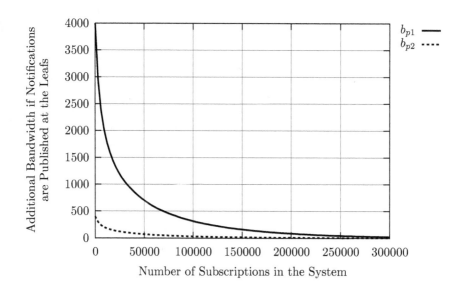

Figure 3.17: Additional bandwidth if notifications are published at the leafs (for $b_{p1} : \omega = 1000$ s^{-1} and $b_{p2} : \omega = 100$ s^{-1})

Overall Bandwidth Saved

The bandwidth used for control traffic b_c, i.e., the bandwidth used for toggling and leasing is

$$b_c = b_o + b_l \qquad (3.33)$$

Hence, the overall bandwidth b saved when applying filtering instead of flooding equals

$$b = b_s - b_p - b_c \qquad (3.34)$$

Figure 3.18 illustrates the behavior of the system in all eight scenarios described in Section 3.5.1 and matches the results from the simulations conducted in Section 3.3.3. The plots of b_{s1} and b_{s2} show the number of messages saved when applying filtering instead of flooding for different publication rates. The other functions represent the control traffic generated by filtering for different expected subscription lifetimes and leasing period combinations. As one can easily see, filtering is a lot more efficient than flooding for smaller numbers of subscriptions. The break-even points, where flooding becomes more efficient than filtering, are the intersections of the control traffic curves $(b_{l\{1|2\}} + b_{o\{1|2\}})$ with the saved traffic curves $(b_{s1/2} - b_{n1/2})$.

In the exemplary setting, b becomes negative if $\mathfrak{N}$ exceeds $299, 503$ subscriptions and b is bounded below by $-b_l$ which equals 2000 messages per second. Hence, in the exemplary setting, filtering is at most $1/60$ worse than flooding (Figure 3.18). Without leasing, filtering would be at least as good as flooding in this case. However, this may not hold for all other scenarios.

Comparing Figure 3.18 with Figure 3.3 on page 52, one can see that the results are very similar. Thus, the analysis confirms the findings of the simulation study carried out in the first part of this chapter.

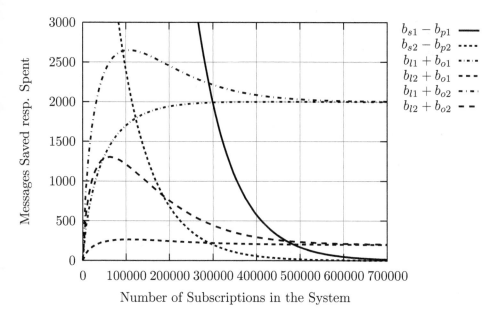

Figure 3.18: Notification messages saved when applying filtering instead of flooding (for $b_{s1} : \omega = 1000$ s^{-1} and $b_{s2} : \omega = 100$ s^{-1}) and control traffic created by filtering and leasing ($b_{l1} : \rho = 60$ s, $b_{l2} : \rho = 600$ s, $b_{o1} : \mu = (600$ s$)^{-1}$ and $b_{o2} : \mu = (60$ s$)^{-1}$)

3.5.4 Generalization

The analysis so far was based on several restrictive assumptions. In this section, a generalization of the analysis is presented, where most of the restrictions are dropped. The assumptions built upon so far are the following:

(1) The broker topology is a complete m-ary tree.

(2) The subscribers can only attach to the leaf brokers.

(3) The interarrival rates for subscriptions and their death rates are equal for all brokers and filter classes.

(4) The publication rates for all filter classes is equal.

(5) The publishers attach either only to the root broker or only to the leaf brokers.

(6) The interarrival times and lifetimes of subscriptions form Poisson processes.

(7) Hierarchical routing is applied.

(8) Identity-based routing is used.

In the following, all assumptions are relaxed except for the last three. Relaxing Assumptions 1 to 5 means that the analysis as it has been done before using closed formulas is not possible anymore. Thus, a formalism based on recursive formulas is proposed. Relaxing Assumption 6 would require a shift from $M/M/\infty$ to $G/M/\infty$ or $G/G/\infty$ queueing systems, changing the basis of the analysis thereby. While this is possible, it will probably not gain any new major insights into the basic interrelationships between the various parameters. Dropping Assumption 7 would change the way subscriptions are forwarded, since (un)subscriptions and toggle messages are then not guaranteed to be forwarded to the root broker only. To be able to drop Assumption 8, additional information, for example, regarding the distribution of subscriptions and notifications would be required. The relaxation of these assumptions is discussed in greater detail in Section 3.5.5.

The first assumption to be relaxed is the assumption that the broker topology is a complete tree. Relaxing it has a severe impact on the calculation of all message complexities. In the following, every broker, thus, has its own set of child brokers $\mathcal{C}(B)$ which is independent of $\mathcal{C}(B')$ of any other broker B'.

Since there is no restriction to closed formulas, it is possible to treat each broker individually and drop any restriction on where the subscribers can attach to (Assumption 2). Therefore, $p_0^f(B)$ as the probability that broker B has no *local* subscription for one filter class f and $\lambda^f(B)$ as the birth rate of subscriptions at broker B for this filter class f are introduced. It can be determined with the subscription death rate $\mu^f(B)$ of B using

$$p_0^f(B) = e^{-\lambda^f(B)/\mu^f(B)} \qquad (3.35)$$

This way, it is possible to drop Assumption 3.

Furthermore, a publication rate for individual filters is introduced, where ω^f is the publication rate of filter f, relaxing Assumption 4 thereby. Moreover, a publication rate $\omega^f(B)$ is introduced for each broker and filter, modeling the case that a publisher for a filter class f can connect to arbitrary brokers. Doing this, Assumption 5 is dropped. The only requirement is, that the publication rates of all brokers for one filter class are consistent with the total publication rate for this filter class:

$$\omega^f = \sum_{B \in \mathcal{B}} \omega^f \qquad (3.36)$$

In the following, the notification and control traffic is determined as done in the restricted scenario discussed so far.

Notification Traffic

Analogous to the previous approach, the case, where all publishers are connected to the root broker, is the starting point and the additional traffic that occurs when publishers can connect to arbitrary brokers is discussed later.

Publisher Connected to the Root. Let $x^f(B)$ be the sum of the occupied routing entries for filter class f at all brokers in the subtree rooted in B. To calculate x^f for arbitrary trees, a recursive formula is proposed. Let $\mathcal{C}(B)$ be the set of all child brokers of broker B and $P_0^f(B)$ be the probability that B is in state 0 for one filter class f, i.e., it has no subscription (neither local nor remote) for this filter class. Then, $P_0^f(B)$ directly depends on $p_0^f(B)$ and on the value of P_0^f of all child brokers of B (if B is not a leaf broker):

$$P_0^f(B) = p_0^f(B) \cdot \prod_{C \in \mathcal{C}(B)} P_0^f(C) \tag{3.37}$$

A broker B has one remote routing table entry for a filter class f and a child broker B' if B' has either a local subscriber or at least one remote routing entry for f. The expected number of remote routing entries in the subtree rooted in B for one filter class f is given by $x^f(B)$ and accordingly depends on the value of P_0^f and x^f of all child brokers:

$$
\begin{aligned}
x^f(B) &= \underbrace{\left(\sum_{C \in \mathcal{C}(B)} (1 - P_0^f(C)) \right)}_{(a)} + \underbrace{\left(\sum_{C \in \mathcal{C}(B)} x^f(C) \right)}_{(b)} \\
&= \sum_{C \in \mathcal{C}(B)} \left(1 - P_0^f(C) + x^f(C) \right) \\
&= |\mathcal{C}(B)| - \sum_{C \in \mathcal{C}(B)} \left(P_0^f(C) - x^f(C) \right)
\end{aligned} \tag{3.38}
$$

The number of occupied remote routing entries $x^f(B)$ for filter class f in the subtree routed in broker B consists of (a) the expected number of remote routing entries in each child broker of B plus (b) the number of remote routing entries in the subtrees rooted in each child broker of B. Starting the recursion with the root broker R the expected number of remote routing entries $x^f(R)$ in the whole system is obtained. In the following, it is set:

$$x^f = x^f(R) \tag{3.39}$$

For complete trees, the recursive equation equals the closed formula in Equation 3.22 when multiplied with the number of filter classes z. The total number of notifications sent is obtained by multiplying the number of remote routing entries for filter class f with the respective publication rate and summing up the results for all filter classes in $\mathcal{F}$ as follows:

$$b'_n = \sum_{f \in \mathcal{F}} \omega^f \cdot x^f \tag{3.40}$$

Publishers Connected to Arbitrary Brokers. As in Equation 3.40, the notification traffic b'_n is obtained by summing up the product of the publication rate and the number of remote routing entries for all filter classes. However, this is only true for the case, where all notifications are published at the root broker. If publishers can connect to arbitrary brokers, the number of additional messages sent due to hierarchical routing has to be added. Additional messages are produced when notifications are sent from the publisher hosting broker towards the root broker if there is no subscriber attached to a broker in the subtree rooted in the forwarding broker.

A notification forwarded by broker B to its parent broker B' is *additional* if there is no subscriber in the subtree rooted in B since it would not be forwarded over this link from B' to B if the notification would have been published at the root broker R. The probability P_0^f that a broker has no routing entry for f has already been determined in Equation 3.37. Hence, it is possible to apply a similar reasoning as for Equation 3.32 and sum up the expected number of additional messages b'_p produced by all brokers which can be determined per broker B and filter class f by multiplying the sum of the probabilities $P_0^f(B')$ of the brokers on the path to the root. $\mathcal{P}(B)$ is introduced as the set of brokers on the path from B to R, including B and excluding R

$$\mathcal{P}(B) = \{B'|B' \neq R \wedge (B' = B \vee B' \in \text{ancestors}(B))\} \tag{3.41}$$

Each $B' \in \mathcal{P}(B)$ has a publication rate $w^f(B)$. The number of additional messages created in the subtree rooted in B for filter f is given by $b_p^f(B)$ as follows:

$$b_p^f(B) = \sum_{C \in \mathcal{C}(B)} \left(w^f(C) \cdot \sum_{B' \in \mathcal{P}(C)} P_0^f(B') \right) \tag{3.42}$$

The number of additional messages b'_p for one filter class f is, thus, given by

$$b_p^f = b_p^f(R) \tag{3.43}$$

The total number of additional messages is the sum of the additional messages b_p^f for all filter classes:

$$b'_p = \sum_{f \in \mathcal{F}} b_p^f \tag{3.44}$$

For the specific case, where all publishers are connected to leaf brokers in a complete m-ary tree, $w^f(B) = 0$ holds for all inner brokers and the total publication rate over all filter classes equals w. Thus, the same result is obtained as in

Equation 3.32:

$$
\begin{aligned}
b'_p &= \sum_{f \in \mathcal{F}} \sum_{B \in \mathcal{B}} \omega^f(B) \cdot \underbrace{\sum_{B' \in \mathcal{P}(B)} P_0^f(B')}_{= \sum_{i=0}^{k-2} p_0^{m^i}} \\
&= \sum_{i=0}^{k-2} p_0^{m^i} \cdot \underbrace{\sum_{f \in \mathcal{F}} \sum_{B \in \mathcal{B}} \omega^f(B)}_{=\omega} = \omega \cdot \sum_{i=0}^{k-2} p_0^{m^i}
\end{aligned}
\tag{3.45}
$$

Notification Traffic Saved. The notification traffic caused by flooding can be obtained analogous to Equation 3.20 by summing the product of the number of links l' and the publication rate for filter class f over all filter classes $\mathcal{F}$:

$$
b'_f = \sum_{f \in \mathcal{F}} \omega^f \cdot l'
\tag{3.46}
$$

For complete trees it is easy to calculate the number of links l' in the hierarchy. For arbitrary tree topologies, l' can be derived using the following recursive formula starting at the root broker R:

$$
l'(B) = \begin{cases} 0 & , \; B \text{ is a leaf} \\ |\mathcal{C}(B)| + \sum_{C \in \mathcal{C}(B)} l'(C) & , \; \text{otherwise} \end{cases}
\tag{3.47}
$$

In the following, it is set

$$
l' = l'(R)
\tag{3.48}
$$

As in Equation 3.25, the notification traffic saved compared to flooding when applying filtering is

$$
b'_s = b'_f - b'_n + b'_p = \left(\sum_{f \in \mathcal{F}} \omega^f \cdot (l' - x^f) \right) + b'_p
\tag{3.49}
$$

Control Traffic

The control traffic b'_c consists of the bandwidth used for leasing b'_l and the bandwidth used for toggling b'_o.

Leasing Traffic. Since each remote routing entry must be refreshed once in a refresh period the resulting leasing traffic b'_l can be determined analogous to the traffic determined in Equation 3.26 as follows:

$$b'_l = \sum_{f \in \mathcal{F}} \rho^{-1} \cdot x^f \tag{3.50}$$

Toggle Traffic. The bandwidth needed for toggling depends on the toggle rate of each broker: if a broker toggles, it sends a toggle message upwards to its parent. The toggle rate $m_o^f(B)$ of a broker B depends on its individual toggle rate m_o^f determined by its local clients and the toggle rate of its child brokers $\mathcal{C}(B)$. In Equation 3.29, the toggle rate $m_o^f(B_l)$ has been obtained for one filter class f and a leaf broker B_l based on the arrival rate λ and the probability p_0 that this broker has no local subscriptions. Subscribers were allowed to only connect to leaf brokers such that it was straightforward to obtain the toggle rate in the system. Now, subscribers are allowed to attach to any broker in the system. Thus, the arrival rate at an inner broker depends on the arrival rate at the broker itself (given by $\lambda^f(B)$) and the toggle rate of all of its child brokers. The arrivals of subscriptions are a Poisson process and the arrival rates of the superposition of multiple Poisson processes is also a Poisson process, where the arrival rate equals the sum of the arrival rates of the particular Poisson processes. Hence, the following recursive formula for the accumulated arrival rate $\lambda_a^f(B)$ of broker B and filter class f is obtained:

$$\lambda_a^f(B) = \begin{cases} \lambda^f(B) & , \quad B \text{ is a leaf} \\ \lambda^f(B) + \sum_{C \in \mathcal{C}} \lambda_a^f(C) & , \quad \text{otherwise} \end{cases} \tag{3.51}$$

With $\lambda_a^f(B)$ it is possible to calculate the toggle rate $M_o^f(B)$ of each broker in the system according to Equation 3.29 as follows:

$$M_o^f(B) = 2 \cdot \lambda_a^f(B) \cdot P_0^f(B) \tag{3.52}$$

Every toggle produces exactly one message. The total number b'_o of toggle messages sent in the system, thus, equals the sum of the toggle rates of all brokers $\mathcal{B}$ for all filter classes:

$$b'_o = \sum_{f \in \mathcal{F}} \sum_{B \in \mathcal{B}} M_o^f(B) \tag{3.53}$$

The results obtained match the results gained for the case, where subscribers can only attach to leaf brokers. In this case, λ and p_0 is equal for all leaf brokers and $p_0 = 1$ and $\lambda = 0$ for all inner brokers. These values are equal for all z filter classes. The toggle rate of broker B on the i-th level is equal to twice the sum of the arrival rates of all leaf brokers that are ancestors of B (i.e., $\lambda \cdot m^{k-1-i}$) times $P_0^f(B)$ with $P_0^f(B) = \prod_{C \in \mathcal{C}(B)} P_0^f(C)$ and $P_0^f(B) = p_0(B)$ if B is a leaf.

Since there are m^i brokers on the i-th level in the tree, the number of toggles in the example is

$$
\begin{aligned}
b'_o &= z \cdot \sum_{B \in \mathcal{B}} M_o^f(B) = 2 \cdot z \cdot \sum_{i=1}^{k-1} m^i \cdot m^{k-1-i} \cdot \lambda \cdot p_0^{m^{k-1-i}} \tag{3.54} \\
&= 2 \cdot \lambda \cdot z \cdot m^{k-1} \cdot \sum_{i=0}^{k-2} p_0^{m^i} = 2 \cdot \lambda \cdot z \cdot \underbrace{m^{k-1} \cdot p_0}_{=\mathcal{N}_L} \cdot \sum_{i=0}^{k-2} p_0^{m^i - 1} \tag{3.55} \\
&= \underbrace{2 \cdot \lambda \cdot z \cdot \mathcal{N}_L \cdot p_0}_{=m_o} \cdot \underbrace{\sum_{i=0}^{k-2} p_0^{m^i - 1}}_{=o} \tag{3.56}
\end{aligned}
$$

This result matches with those of Equation 3.27 and 3.30.

Messages Saved

The overall number of messages saved when applying filtering instead of flooding b' according to Equation 3.21 is

$$
\begin{aligned}
b' &= b_f - b'_n + b'_p - b'_l - b'_o = b'_s - b'_l - b'_o \\
&= \left(\sum_{f \in \mathcal{F}} \left(\omega^f \cdot (l' - x^f) + b^f_p \right) \right) - \rho^{-1} \cdot \left(\sum_{f \in \mathcal{F}} x^f \right) - \left(\sum_{f \in \mathcal{F}} \sum_{B \in \mathcal{B}} M^f_o(B) \right) \\
&= \sum_{f \in \mathcal{F}} \left(\omega^f \cdot (l' - x^f) + b^f_p - \rho^{-1} \cdot x^f - \sum_{B \in \mathcal{B}} M^f_o(B) \right)
\end{aligned}
\tag{3.57}
$$

3.5.5 Discussion

In the initial analysis, the work presented based on several restrictive assumptions which made it possible to provide a formalism consisting of closed formulas. In the generalization, several of those assumptions were dropped and, thus, the class of scenarios which are covered by the analysis were broadened as subscribers can now attach to every broker in the system. Moreover, it is possible to set the birthrate and the lifetime of subscriptions individually for each broker and filter class as well as the publication rate of connected publishers. However, the assumption still holds that the arrivals and lifetimes of subscriptions are a Poisson process and determine $P^f_0(B)$ accordingly. This is the reason why it is possible to simply add the arrival rates of parent and child brokers in order to obtain the accumulated arrival rate of the superpositioned processes at a broker. Using another probability distribution than the exponential distribution for the arrivals affects the calculation of the number of toggle messages in the system. The extension for other probability distributions are left open for future work.

Another advantage of the generalization is that it is now possible to examine the effects of locality by parameterizing the probability distribution on each broker separately. This way, it is possible to model, for example, local clusters of identical subscriptions. Before having this formalism, it was necessary to conduct simulations to analyze the behavior of the system in such complex settings.

The generalized analysis is still based on hierarchical and identity-based routing. It would be interesting to consider other routing strategies like peer-to-peer routing instead of hierarchical routing. This is possible with the formalism presented. However, it is necessary to consider the fact that forwarding of (un)subscriptions and toggle messages is not limited to the root broker. Thus, the message complexity is expected to increase in this case. In order to use more sophisticated routing algorithms than identity-based routing (like covering-based routing), more changes are required. Although not straightforward it is possible to incorporate such routing algorithms into the analytical framework presented. In this case, it is necessary to additionally model the covering relations between the filters different subscriptions carry. If the probability would be known that a subscription covers another one, it would be possible to incorporate this into the formalism given above since it would be possible then to reason about the probabilities that subscriptions and toggle messages are forwarded. This issue is left for future work.

3.6 Related Work

In the following, related work is discussed in the areas of self-stabilization, fault tolerance, and analysis of publish/subscribe systems and put in the context of the work presented in this chapter.

Self-Stabilizing Content-Based Routing

In the field of self-stabilizing routing in publish/subscribe systems the closest related work is by Shen and Tirthapura [142]. In their approach, all pairs of neighboring brokers periodically exchange "sketches" of those parts of their routing tables concerning their other neighbors to detect corruption. Those sketches exchanged are lossy because they are based on Bloom filters (which are a generalization of hash functions) [25]. However, due to the information loss it is not guaranteed that an existing corruption is detected deterministically. Hence, the algorithm is not self-stabilizing in the usual sense. Moreover, although generally all data structures can be corrupted arbitrarily, the authors' algorithm computes the Bloom filters incrementally [143]. Thus, once a Bloom filter is corrupted, it may never be corrected, resulting in a system which may never stabilize! Furthermore, clients do not renew their subscriptions. Without this, corrupted routing entries regarding local clients are never corrected. Finally, in its current form, their algorithm is restricted to identity-based routing.

An interesting idea is that of limiting the message complexity with transient link failures. However, the solution proposed by the authors relies on a topology maintenance algorithm which actively informs the self-stabilizing routing layer about failed links. In addition to that, the algorithm needs to know the size of the subtree that is rooted in the broker it runs on. The authors neither explain how they obtain this information nor do they discuss the effect on the self-stabilizing behavior of the whole system.

The only other directly related work realizes self-stabilizing publish/subscribe on top of a peer-to-peer network [159]. Therefore, the authors add self-stabilizing mechanisms to an earlier work by Datta et al. which supports topic-based publish/subscribe by maintaining one overlay tree ("layer") per topic [54]. The authors claim that they also support content-based routing. However, this support is rather restricted as it still requires topics and super-peers to which clients connect according to the topics they want to subscribe for. It, thus, does not conform to the general notion of content-based routing which does not require topics.

Fault-Tolerant Publish/Subscribe Systems

Strom and Jin propose using a stateful approach for publish/subscribe systems [87, 146]. They introduce a history of events held in the system, enabling new functionality thereby (e.g., publishing an hourly mean value of notification values). This way, it is also possible to handle transient faults and compensate lost notifications. The authors consider transient faults regarding links or notifications which may get lost. However, they do not handle arbitrary perturbations, e.g., of routing tables. Thus, the system they propose is not self-stabilizing.

There has been some work on layering a publish/subscribe system on top of a peer-to-peer routing substrate like PASTRY [136] (the HERMES publish/subscribe system [128, 129]), CAN [133] (MEGHDOOT [70]), and CHORD [144] (as described in [148]). These peer-to-peer routing substrates are designed with failing or leaving nodes in mind and are, thus, very robust with respect to these faults. Layering a publish/subscribe system on top of such a routing substrate can be beneficial with respect to faults. However, the fault tolerance mechanisms provided by the routing substrate do only relate to the broker overlay network. The routing tables held by the brokers are not considered such that it is necessary to employ another mechanism if they are corrupted. Aside from this, it is not clear yet whether the routing substrates used are self-stabilizing themselves.

Costa et al. [46, 47, 48] take an interesting approach to obtain reliability by using a probabilistic algorithm which relies on the theory of epidemics [55]. By caching notifications and gossiping cache contents the authors try to reach eventual notification completeness (similar to what Shen and Tirthapura do with subscriptions in [142]). This approach is well suited for highly dynamic environments with link failures and message loss. Corrupted routing tables are not taken into account such that these may lead to an incorrect state of the system from which it may never recover. Moreover, their approach only provides probabilistic delivery guarantees and is not able to provide a given message ordering.

The authors of [22, 146] propose an algorithm which guarantees notification delivery in the GRYPHON publish/subscribe system [145]. It is able to cope with lost or reordered messages by employing an approach which is based on (negative) acknowledgements and event histories. The failure model comprises node failures, dropped and reorders messages, as well as link failures. Publishers log messages in stable storage to be able to resend them. However, arbitrary corruption of data structures like routing table entries is not considered and not handled.

Other work in this area comprises the use of virtual time vectors in combination with redundant paths [164]. This way, the authors are able to provide complete and in-order delivery of messages. The authors of [107] take a similar approach and try to render the publish/subscribe system tolerant with respect to node and link failures by using a replication algorithm. In contrast to the fault model used here, the fault model of both solutions does not consider perturbations of routing tables and is not self-stabilizing.

Self-Stabilizing Multicast Communication

Recently, there have been some research efforts for implementing self-stabilizing multicast communication in mobile ad hoc networks [71, 86]. They aim at building a self-stabilizing spanning tree which is well suited to the application scenario. Several algorithms have been proposed for self-stabilizing spanning trees like the ones by Dolev et al. [59], and Aggarwal and Kutten [4], just to mention two of them. Gärtner gives a good overview of self-stabilizing spanning tree algorithms in [66].

Self-stabilizing spanning tree algorithms are also an interesting way to maintain the broker network, the publish/subscribe routing is layered upon. However, since content-based routing is far more complex than multicast communication, simply maintaining a spanning tree does not help in making content-based routing self-stabilizing. In [120], the authors describe an algorithm which exploits IP multicast for content-based routing in publish/subscribe systems. Therefore, the algorithm maps subscriptions more or less accurate to multicast groups. Although single multicast groups may be realized with a self-stabilizing spanning tree algorithm, the whole system is not self-stabilizing since the group membership management, for example, is generally not self-stabilizing.

Analysis of Publish/Subscribe Systems

At the time writing, there are only three publications which deal with analyzing publish/subscribe systems. The first one by Bricconi et al. [27] presents a rather simple model with a lot of restrictions for the analysis of the JEDI publish/subscribe systems. The model is mainly used to calculate the number of events received by each broker using a uniform distribution of subscriptions. To model the multicast communication, the authors introduce a spreading coefficient between 0 and 1 which models the probability that a broker in a given distance (in hops) of the publishing broker receives an event published.

The other analysis has been published by Baldoni et al. [13]. Besides a system model, the authors propose an analytical model. Instead of analyzing the message complexity as done in Section 3.5, the proposed model is only used to calculate the number of notifications that are missed by subscribers due to network delays.

There is only one other analytical model that deals with questions regarding message complexity of publish/subscribe systems. In this recent work the same assumptions are made with respect to the broker topology as in the initial analysis [37]. The distribution of subscribers and publishers is assumed to be uniform such that mean values can be used to make the analysis easier. However, the authors take a different approach in the routing protocols considered which makes it hard to compare it to our approach. Nevertheless, the generalization of the approach presented in this chapter is far more powerful since it is easily possible to model locality and arbitrary individual parameterizations of subscription distributions and their respective lifetimes. It is important to note that this approach was published more than two years after the first results presented here had been published [82].

3.7 Discussion

In this chapter, the notion of self-stabilizing content-based routing for publish/subscribe systems has been introduced. Self-stabilizing algorithms for simple, identity-based, covering-based, and merging-based routing have been proposed. It has been discussed that the self-stabilizing content-based routing algorithms presented induce extra costs in form of additional messages. It has also been shown in an experimental evaluation in which scenarios self-stabilizing hierarchical identity-based routing is superior to flooding. The simulations assumed a rather conservative setting with a uniform distribution of subscriptions at the leaf brokers; thus, they mark a rather worst-case scenario. Furthermore, several generalizations and extensions have been proposed: a wrapper algorithm which can be used to make arbitrary correct routing algorithms self-stabilizing, the integration of advertisements into self-stabilizing routing, and the combination with (a)cyclic peer-to-peer routing.

Self-stabilization has been realized by a periodic task which "pushes" the system towards a correct state. The cost of this periodic task may not pay off in all situations. In small-scale energy-sensible scenarios like sensor networks or mobile ad hoc networks, for example, one has to take care that the self-stabilizing mechanism does not lead to an energy shortage due to a large message overhead which wastes all the energy available. However, especially in those scenarios, self-stabilization is a property of great value. In [154], a self-stabilizing routing algorithm has been applied as presented in Section 3.3 to create a self-stabilizing role-assignment mechanism in actuator/sensor networks. Similar to small-scale networks, the extra load imposed on large-scale networks may not justify the advantages of self-stabilization. Here, it could be of great value to segment the network into different "zones" with different stabilization demands and, hence, different message overhead due to self-stabilization. An adaptive approach would be the next step—however, it has to be taken care that the self-stabilization guarantees still hold since adaptive solutions often rely on complex soft state. More complex soft state with more complex interdependencies inside the state makes it increasingly difficult to ensure convergence from an incorrect to a correct state without external intervention. This issue is left for future work.

In the second part of this chapter, a stochastic analysis of the message complexity in publish/subscribe systems has been presented which allows for the first time to reason about the message overhead of a publish/subscribe system without having to rely on simulations. The analysis confirmed the findings from the simulation study in the first part. For this scenario, closed formulas have been proposed for complete trees and hierarchical routing. Later, the analysis was generalized to arbitrary trees, where publishers and subscribers can connect to any broker in the system. The clients' (i.e., subscribers' and publishers') behavior can be parameterized per broker and is, thus, able to capture a wide range of scenarios. Given the formalism, it is now possible to analyze the message complexity of self-stabilizing publish/subscribe routing without having to rely on long-running simulations, where implementation details, which have an impact on the results, are often not clear to the reader.

For the analysis, hierarchical identity-based routing was required and more complex routing algorithms like covering-based and merging-based routing have not been considered. This is due to the fact, that handling these routing algorithms requires additional assumptions and models regarding the covering- or merging-probability of subscriptions and their distribution. It is highly probable the it is possible to include both into the analytical framework presented. This topic is left open for future work. Furthermore, it is assumed that subscription arrivals can be modeled as Poisson processes and that subscription lifetimes are exponentially distributed. This may not hold in all scenarios. Incorporating other probability distributions into the analysis is another issue which is left open for future work.

The analysis provides an important foundation for modeling and analyzing publish/subscribe systems. Providing tools to formally analyze publish/subscribe systems is an important contribution in an area, where evaluations mostly rely on extensive simulation studies and often neither the simulation code nor the underlying datasets are open to the reader.

4 Reconfiguring Broker Overlay Networks

Contents

4.1 Introduction

Publish/subscribe systems are well suited to and often used in dynamic scenarios, where system parameters may vary significantly in an unexpected and unpredictable way. Examples for large-scale and small-scale scenarios, respectively, are Internet-wide information dissemination systems and e-home applications with spontaneously joining and leaving nodes. In a distributed implementation, brokers cooperate in a peer-to-peer fashion and contribute on their part to the infrastructure they use. When new nodes join the system or participating nodes leave, for example, it might be beneficial or even necessary to reconfigure the broker overlay network topology in order to keep the network connected or to optimize its performance. Thus, support for reconfiguring the broker overlay topology of a publish/subscribe system is essential for managing or even enabling the evolution of the broker topology of a publish/subscribe system in a dynamic environment.

In this chapter, the problem of reconfiguring the broker overlay network of conventional and self-stabilizing publish/subscribe systems is discussed. In the beginning, faults are not discussed and it is assumed that reconfigurations are carried out in a fault-free scenario. The chapter starts with presenting the notion of reconfiguration used in this book in Section 4.2 which is assumed to be planned and executed as part of system management. Elementary reconfigurations are introduced which can be combined to execute complex reconfigurations. Subsequently, the general challenges of reconfiguring publish/subscribe overlay topologies are discussed in Section 4.3. This also comprises the integration of reconfigurations into the publish/subscribe model and the identification of guarantees which are valuable to provide during reconfiguration.

A novel algorithm is presented for managed reconfigurations in controlled environments for conventional publish/subscribe systems in Section 4.4. It is the first algorithm which is capable of guaranteeing FIFO-publisher as well as causal message ordering when carrying out reconfigurations. Besides message ordering, the focus is on preventing message loss and minimizing message overhead. The influence of the particular ordering requirements and the reconfiguration scenarios on the performance of the algorithm are examined in a simulation study. The results are compared to those of the classic STRAWMAN approach which is often cited in literature.

In Section 4.5, reconfigurations in self-stabilizing publish/subscribe systems are considered which apply self-stabilizing content-based routing as presented in Chapter 3. For these systems, it is not possible to apply the algorithm presented for regular publish/subscribe systems because self-stabilization implies subtle issues which have to be handled appropriately. First, the self-stabilizing content-based routing layer is complemented with a self-stabilizing broker overlay network in order to realize a self-stabilizing publish/subscribe system. Both layers need to be coordinated in order to prevent service interruptions due to reconfigurations. Therefore, a coloring mechanism is introduced which coordinates actions on both layers. The result is a complete self-stabilizing publish/subscribe stack which is capable of implementing reconfigurations without message loss while maintaining message ordering. This chapter concludes with an overview of related work (Section 4.6) and a discussion of the results (Section 4.7).

The ability to reconfigure the broker topology of a publish/subscribe system is an essential prerequisite for the next chapter which focuses on the problem of how to enable a publish/subscribe system to optimize its broker overlay topology autonomously. The optimizations found have to be carried out, for example, by using the algorithms presented in this chapter.

4.2 Reconfigurations

While faults happen suddenly and may lead to abrupt changes in the broker topology, a reconfiguration is considered here as a *cooperative* process which is usually planned in advance and tolerates a delay to take effect.

In the following, different types of reconfigurations of the broker overlay topology are presented. Then, all possible reconfigurations are reduced to three basic reconfigurations which are called *elementary reconfigurations*. It will be shown that arbitrary reconfigurations can be carried out with a sequence of elementary reconfigurations and the notion of reconfiguration will be defined accordingly.

4.2.1 Types of Reconfiguration

Four different types of reconfiguration of the broker overlay network and their impact on the maintenance of the affected brokers' routing tables are discussed in the following.

Broker Addition

When a new broker B wants to join the overlay network, it has to connect to at least one broker while maintaining the acyclic structure of the topology at the same time. If B shall be added as a leaf broker, this is easily accomplished by connecting it to one broker and then starting it. The integration is complete when B has obtained *all* necessary subscriptions (and advertisements) from its neighbor broker. After that, B is part of the broker overlay network and integrated into the publish/subscribe system.

Adding B as an inner broker is far more complicated. In this case, it is necessary to remove at least one link from the topology to keep the topology acyclic after integrating B. To finalize its integration, B needs to make its routing table consistent with the routing tables of all of its neighbor brokers. Additionally, the removal and addition of links may require other brokers to update their routing tables.

Broker Removal

The problem of removing a broker B is similar to that of adding a broker. If B is a leaf broker it simply has to cancel all its subscriptions (and advertisements) and can be disconnected subsequently. If B is an inner broker, links have to be relocated and routing tables have to be synchronized which affects at least those brokers to which B is connected.

Link Addition

As already discussed broker addition or removal may require adding an additional link to the topology. Since it is necessary to keep the topology acyclic, a newly inserted link (which does not connect a new leaf broker with the rest of the topology) automatically results in the removal of another link. This link has to be chosen from the links forming the cycle which results in the topology when inserting the new link. This path will be called the *reconfiguration cycle* in the following. Figure 4.1 depicts an example of a reconfiguration cycle, where the link $\overline{B_1^a B_2^a}$ (called a in the following) shall be inserted into the topology. It replaces the link $\overline{B_1^r B_2^r}$ (called r in the following). In this example, the reconfiguration cycle consists of the brokers B_1^a, B_1, B_1^r, B_2^r, B_2, and B_2^a.

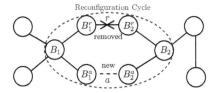

Figure 4.1: Example of a reconfiguration cycle (the dashed link is inserted)

Since the contents of the routing tables is strongly tied to the broker topology, adding a new link as in Figure 4.1 may require an update of the routing tables of other brokers which are not an endpoint of the new link or the link removed for it. For example, B_1 in Figure 4.1 may need to forward notifications for which only B_2 subscribed after the reconfiguration in direction of B_1^a while the same notifications would have been forwarded towards B_1^r before.

Link Removal

Analogous to inserting a new link, removing a link requires a new link to be added in order to keep the topology connected. The cycle that results from adding both links to the topology is also called reconfiguration cycle as introduced above. Thus, adding or removing a link finally comes down to the exchange of one link with another one.

4.2.2 Elementary Reconfigurations

In the previous section, four different reconfigurations of the broker topology have been identified: link exchange, and inner and outer broker removal and addition. These reconfigurations can also be carried out by a sequence of only three *elementary reconfigurations*. In the following, they are described and it is shown how an arbitrary complex reconfiguration can be constructed with a sequence of them.

Adding a Leaf Broker

As described in Section 4.2.1, adding a leaf broker is easy since it implies only adding one link. Moreover, it is only necessary to make the routing table of the new leaf broker consistent with its neighbor broker. Thus, adding a leaf broker is a relatively "cheap" operation. Figure 4.2(a) depicts an example scenario, where broker B^a is added to broker B using the new link a.

Removing a Leaf Broker

Similar to adding a new leaf broker it is also easy to remove one. On the topology level this only requires tearing down one link, while on the routing layer, the leaf broker to be removed must remove all its subscriptions (and advertisements) by issuing respective unsubscriptions (and unadvertisements).

Figure 4.2(b) shows an example scenario, where broker B^r should be removed which implies the removal of the link r which connects B^r with its sole neighbor B.

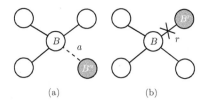

(a) (b)

Figure 4.2: Adding and removing a leaf broker

Link Exchange

This type of reconfiguration encompasses the action of replacing a link in the topology with another one. When removing (adding) a link, it is necessary to add (remove) another link to keep the topology acyclic and connected. An example is depicted in Figure 4.3, where link r is replaced with link a. With a, the topology is connected and acyclic again if r is removed.

The problem lies in making the routing tables of the brokers consistent with the new topology structure. Messages, which have been routed over r before the reconfiguration, have to be routed over a afterwards. Later it will become clear that the routing tables of all brokers on the reconfiguration cycle may need to be updated. Thus, exchanging links in the topology may require considerable effort.

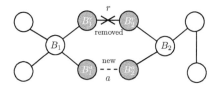

Figure 4.3: Exchanging one link with another one

4.2.3 Complex Reconfigurations

Reconfigurations that cannot be implemented with one elementary reconfiguration step are called *complex reconfigurations* in the following. They include adding and removing inner brokers as discussed previously but also transforming the topology from one into another arbitrary spanning tree.

One can prove that complex reconfigurations can be carried out by a sequence of elementary reconfigurations. For example, adding an inner broker B can be achieved with the following procedure: add B as a leaf broker and execute a sequence of link exchanges until B has reached its final position, i.e., until the tree has a topology, where B resides at the intended position. For removing an inner broker B, it is required to execute a set of link exchanges until B is a leaf broker and the rest of the brokers in the tree are connected as intended. Then, it is easy to remove B as already described.

Before proving that complex reconfigurations can be composed by elementary reconfigurations, the notion of *distance* between two spanning trees $\mathfrak{T}_1$ and $\mathfrak{T}_2$ for an identical set of nodes is introduced which represents the number of edges contained in $\mathfrak{T}_2$ but not in $\mathfrak{T}_1$.

Definition 7 (Distance). *Let $\mathfrak{T}_1 = (\mathcal{V}, \mathcal{E}_1)$ and $\mathfrak{T}_2 = (\mathcal{V}, \mathcal{E}_2)$ be spanning trees of the connected graph $G = (\mathcal{V}, \mathcal{E})$. The distance $D(\mathfrak{T}_1, \mathfrak{T}_2)$ of $\mathfrak{T}_1$ to $\mathfrak{T}_2$ is given by the number of edges that are present in $\mathcal{E}_2$ but not in $\mathcal{E}_1$*

$$D(\mathfrak{T}_1, \mathfrak{T}_2) := |\{e| e \notin \mathcal{E}_1 \wedge e \in \mathcal{E}_2\}|$$

Please note that $\mathfrak{T}_1$ equals $\mathfrak{T}_2$ if $D(\mathfrak{T}_1, \mathfrak{T}_2) = 0$ holds and that D is symmetric:

Lemma 2 (Symmetry). *The relation D is symmetric*

$$D(\mathfrak{T}_1, \mathfrak{T}_2) = D(\mathfrak{T}_2, \mathfrak{T}_1)$$

Proof. Assume that $D(\mathfrak{T}_2, \mathfrak{T}_1) > D(\mathfrak{T}_1, \mathfrak{T}_2)$ without loss of generality. The total number of edges in $\mathfrak{T}_1$ and $\mathfrak{T}_2$ is equal because both are spanning trees of G

$$|\{e|e \in \mathcal{E}_1\}| = |\{e|e \in \mathcal{E}_2\}| = |\mathcal{V}| - 1 =: l$$

The proof is by contradiction.

$$
\begin{aligned}
D(\mathfrak{T}_2, \mathfrak{T}_1) \;&>\; D(\mathfrak{T}_1, \mathfrak{T}_2) \\
\Leftrightarrow |\{e|e \in \mathcal{E}_1 \wedge e \notin \mathcal{E}_2\}| \;&>\; |\{e|e \in \mathcal{E}_2 \wedge e \notin \mathcal{E}_1\}| \\
\Leftrightarrow l - |\{e|e \in \mathcal{E}_1 \wedge e \in \mathcal{E}_2\}| \;&>\; l - |\{e|e \in \mathcal{E}_2 \wedge e \in \mathcal{E}_1\}| \quad \text{\textlightning}
\end{aligned}
$$

$\square$

Theorem 2. *For a fully connected graph $G = (\mathcal{V}, \mathcal{E})$, a spanning tree $\mathfrak{T}_1 = (\mathcal{V}, \mathcal{E}_1)$ can be transformed into an arbitrary spanning tree $\mathfrak{T}_2 = (\mathcal{V}, \mathcal{E}_2)$ with $\mathfrak{T}_1 \neq \mathfrak{T}_2$ by only a sequence of link exchanges.*

Proof. In the following, step-by-step instructions are provided to transform $\mathfrak{T}_1$ into $\mathfrak{T}_2$. Let

$$
\begin{aligned}
k \;&:=\; D(\mathfrak{T}_1, \mathfrak{T}_2), \\
\mathcal{E}'_1 \;&:=\; \{e|e \in \mathcal{E}_1 \wedge e \notin \mathcal{E}_2\}, \text{ and} \\
\mathcal{E}'_2 \;&:=\; \{e|e \in \mathcal{E}_2 \wedge e \notin \mathcal{E}_1\}.
\end{aligned}
$$

Replace one edge in $\mathcal{E}'_1$ with one from $\mathcal{E}'_2$. Doing this $D(\mathfrak{T}_1, \mathfrak{T}_2)$ is reduced by one. Now repeat this $k - 1$ times, recalculating $\mathcal{E}'_1$ and $\mathcal{E}'_2$ each time. Due to the symmetry of D it is ensured that there is always an edge in $\mathcal{E}'_1$ which can be replaced with one from $\mathcal{E}'_2$. This transforms $\mathfrak{T}_1$ into $\mathfrak{T}_2$ because $D(\mathfrak{T}_1, \mathfrak{T}_2)$ equals 0 afterwards, i.e., $\mathfrak{T}_1 = \mathfrak{T}_2$. Thus, it was possible to transform $\mathfrak{T}_1$ to $\mathfrak{T}_2$ using only k link exchanges. $\square$

Transforming a topology $\mathfrak{T}_1 = (\mathcal{V}_1, \mathcal{E}_1)$ into a topology $\mathfrak{T}_2 = (\mathcal{V}_2, \mathcal{E}_2)$, where $\mathcal{V}_1 \neq \mathcal{V}_2$ can accordingly be accomplished by first adding the set $\{v|v \in \mathcal{V}_2 \wedge v \notin \mathcal{V}_1\}$

of nodes to $\mathfrak{T}_1$ which are contained in $\mathfrak{T}_2$ and not in $\mathfrak{T}_1$ and removing the set $\{v|v \in \mathcal{V}_1 \wedge v \notin \mathcal{V}_2\}$ of nodes which are part of $\mathcal{V}_1$ and not $\mathcal{V}_2$ from $\mathfrak{T}_1$. The resulting topology $\mathfrak{T}_1'$ can afterwards be transformed into $\mathfrak{T}_2$ by exchanging links as explained previously. Inner nodes can be removed from $\mathfrak{T}_1$ by executing link exchanges until the broker is a leaf and can be remove accordingly. New brokers are added as leafs.

The focus in this chapter is on elementary reconfigurations. More precisely, the definition of a reconfiguration used is as follows.

Definition 8 (Reconfiguration). *A reconfiguration is a change of the broker overlay topology, comprising leaf broker removals, leaf broker additions, and link replacements that can be delayed for a finite time. This change must maintain the spanning tree property of the broker overlay topology.*

It is important to note that reconfigurations as defined in Definition 8 do always maintain the spanning tree property of the broker overlay topology. Changes in the topology that lead to a partitioned network or a cyclic graph, for example, are not reconfigurations in this sense and are considered as faults. The definition only considers three different reconfiguration types.

4.3 Challenges

The loose coupling between publishers and subscribers in a publish/subscribe system suits dynamic changes in the system very well: new publishers may join while old publishers leave without any impact on the subscriber side. The same holds vice versa but in the former case, advertisements have to be sent out or cancelled if used, while in the latter case, (un)subscriptions have to be propagated. This is, however, transparent to the clients because this management task is handled by the notification service and, thus, neither the subscribers nor the publishers are normally aware of new or leaving participants in the system. From the clients' perspective this transparency should also hold for reconfigurations inside the notification service, i.e., the clients in general do not expect disturbances like message loss, duplication, or reordering due to reconfigurations of the notification service. This, however, is not easy to accomplish as shown in the following.

The common architecture for publish/subscribe systems which relies on an acyclic broker overlay topology is used here. From the programming perspective, clients (subscribers and publishers) are applications while brokers are started locally on the computer which hosts the clients or may run standalone. In the former case, a leaving node may not only comprise leaving publishers or subscribers but also a leaving broker. In an acyclic topology, this may lead to a partitioned network if the broker is not a leaf in the broker topology. The same is true if a standalone broker without local clients goes off-line (e.g., for maintenance). In this section, the problems that may occur during reconfiguration are explored and the integration of reconfigurations into the publish/subscribe system model is discussed.

4.3.1 Issues During Reconfiguration

For the reconfiguration types discussed so far, the focus is on keeping the topology acyclic and the state of the routing tables consistent with the topology. The rationale behind this is that correct routing table configurations will route messages correctly and, thus, the correctness properties from Definition 5 will hold after a reconfiguration. Since a fault-free environment is assumed, it is reasonable to assume that this also holds before the reconfiguration. However, this is not necessarily true for the time period between the start and the end of a reconfiguration.

It has already been discussed that reconfigurations of the topology of the notification service at runtime are expected to be executed without any service interruption. The service guarantees that should also hold during reconfigurations center around the requirement that reconfigurations have to be transparent to the clients. In the following, two important guarantees are discussed that are often required during reconfiguration: message completeness and message ordering.

Message Completeness

When links are replaced, new routing table entries have to be installed while superfluous routing entries have to be removed in order to keep the routing tables of the brokers on the reconfiguration cycle consistent with the broker network topology. At this point, timing is crucial since new routing table entries are necessary for routing messages over the new link and removing superfluous routing table entries prevents messages from being duplicated and from being routed towards directions, where no subscribers reside. When notifications are forwarded in this time period, they may get lost or be duplicated due to race conditions.

Message completeness is a very important guarantee which inherently relates to the basic functionality of correct publish/subscribe systems (cf. Definition 5). Two types of messages are considered: control messages and notifications. Control messages are used by the routing algorithms and can comprise subscriptions, unsubscriptions, advertisements, and unadvertisements. Please note that neither faults, nor the general feature of message-complete publish/subscribe systems are considered which goes beyond the scope of this book [147].

In the context of reconfigurations, where a broker overlay network is transformed from topology $\mathfrak{T}_1$ to topology $\mathfrak{T}_2$, message loss is defined as follows.

Definition 9 (Message Loss During Reconfiguration). *When a broker overlay network is transformed by a reconfiguration from topology $\mathfrak{T}_1$ to topology $\mathfrak{T}_2$ and provided that no fault occurs, a message m published is called* lost *if one of the following statements holds:*

- *(m is a notification) A client c does not receive m, while c would have received m in $\mathfrak{T}_1$ provided that the topology would not have been reconfigured. To prevent race conditions due to unsubscriptions or unadvertisements, it is required that c does not issue an unsubscription concerning m for a long enough time period.*

- *(m is a control message) The configuration of the brokers' routing tables (advertisement tables) in $\mathfrak{T}_2$ prevents satisfying the liveness condition in Definition 5 during or after the reconfiguration because missing routing (advertisement) table entries lead to lost notifications.*

It is important to note that due to the asynchronous nature of publish/subscribe communication, (un)subscriptions and (un)advertisements become only gradually active. Thus, Definition 9 considers for notifications only those cases, where subscriptions or advertisements are not changing for a long enough time period while reconfiguring the topology. Otherwise it may happen, for example, that due to the new topology it may take longer to forward a notification from a publisher-hosting broker B_p to a subscriber-hosting broker B_s in $\mathfrak{T}_2$ than in $\mathfrak{T}_1$: if B_s issues an unsubscription in $\mathfrak{T}_2$ short before the notification has reached B_s, the notification may have been received in $\mathfrak{T}_1$ but is not received in $\mathfrak{T}_2$ due to differing delays. In this case, the notification would be lost according to Definition 9 which is certainly counter-intuitive.

Regarding reconfigurations, the focus is on the prevention of message loss in the following. There are different reasons why messages might get lost and the consequences are manifold. They are discussed in the following.

Notification Loss. When performing a reconfiguration, a notification may get lost in two situations: (i) the notification is sent over a link which is torn down in this moment, or (ii) the notification arrives at an intermediate broker which drops it since the routing table entries needed for forwarding this notification correctly have not been installed yet. The first case can only be prevented by introducing a synchronization mechanism such that a link is not torn down while a notification is sent over it. This does not pose a problem if it is assumed that reconfigurations can be delayed. The second case represents a race condition which must be prevented by appropriate measures. A more detailed description of race conditions is given below.

The impact of notification loss is depending on the application but should be prevented in general since it violates the liveness property of a correct publish/subscribe system.

Control Message Loss. Similar to notifications, (un)subscriptions or (un)advertisements may be lost due to dropped links, out-dated routing tables, or race conditions. However, while lost notifications represent a transient error, a lost control message may lead to a permanent error and is, thus, more severe. If a client, for example, issues a subscription which is lost during dissemination, it may happen that the client never receives every publication produced that matches this subscription, because the subscription did not reach a respective publisher. This violates the liveness property of a correct publish/subscribe system.

Race Conditions. During reconfiguration, the most difficult part is to avoid race conditions. This issue is, thus, discussed in greater detail in the following. The routing tables in a publish/subscribe system are built according to the broker overlay topology. Hence, they have to be adapted when the topology changes. A simple example is depicted in Figure 4.4, where link r is replaced by link a. The little arrows depict the routing table entries for one filter with one subscriber connected to B_2. In this example, it becomes obvious that a reconfiguration affects the routing tables of brokers B_1^r, B_1, and B_1^a and that a notification forwarded by B_1 to B_1^r in Figure 4.4(a) may be dropped at B_1^r if the topology is reconfigured to the one depicted in Figure 4.4(b) meanwhile (this is due to the fact that routing algorithms normally do not send a notification they received back to the sender). The same applies to subscriptions and advertisements. Similarly, notifications may be duplicated if they are sent over a *and* r. It is, hence, reasonable to coordinate the forwarding of messages with the reconfiguration of the topology *and* the routing tables.

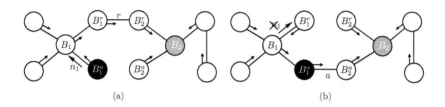

Figure 4.4: Example for message completeness violation

Message Ordering

Besides message loss, a reordering of notifications may happen due to reconfig-urations. Assume the same scenario as in Figure 4.4, where link r is replaced by link a and that before the reconfiguration starts, B_1^a publishes notification n_1, where B_2 has subscribed for a matching filter. After the reconfiguration has been carried out, B_1^a publishes n_2, which then reaches B_2 over the new link as depicted in Figure 4.5. This results in a much shorter path with respect to broker hops. It may, thus, happen that n_2 arrives before n_1 at B_2. In this case, the order in which B_1^a published the notifications is different from the order in which the notifica-tions arrived at B_2. This could not happen in a static topology, because the FIFO property of the communication links between brokers guarantees a FIFO-publisher ordering of notifications received by the subscribers.

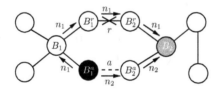

Figure 4.5: Example for message ordering violation

The reason why keeping a message ordering in face of reconfigurations is difficult, although FIFO links are required between the brokers, is that messages might be forwarded over different paths to the same broker because of a reconfiguration.

Due to varying link delays on the way messages may then overtake each other on different routes.

4.3.2 Integration

It is an important question how to integrate a reconfiguration mechanism into a publish/subscribe system. The basic interface operations of a publish/subscribe system are the operations pub(), sub(), and unsub() (and adv() and unadv(), if supported). For an easy integration it is sensible to rely only on these operations. Doing this, it is possible to benefit from the abstractions regarding the routing algorithm used provided by the interface operations when incorporating reconfigurations with different routing algorithms.

However, from the issues explained previously it is obvious that implementing reconfigurations only with theses operations will not suffice since notifications may need to be delayed, queued, or replayed to prevent message loss and guarantee a required ordering. Thus, it is not possible to simply layer the reconfiguration algorithm on top of the content-based routing layer. Therefore, additional coordination mechanisms will be installed inside the publish/subscribe system which are discussed in the following section.

4.3.3 Guarantees

This section focuses on the main challenges already outlined. As stated, reconfigurations are not handled due to faults. Instead the focus is on reconfigurations that change the broker overlay network consistently, for example, by an administrative order (Definition 8).

In the previous section, the most important guarantees when reconfiguring the topology of a publish/subscribe system at runtime have already been described. The example in Figure 4.5 on the previous page illustrated that a reconfiguration can mess up the message ordering. In the following, the guarantees are presented which are to be supported (prevention of message loss, FIFO-publisher ordering, and causal ordering) and others which may be required but are not targeted here.

Prevention of Message Loss

This basic requirement of preventing message loss has already been discussed in length in Section 4.3.1. However, due to timing issues, this guarantee remains a little bit vague in an asynchronous publish/subscribe system although its meaning is easy to understand intuitively.

Ordering

Many applications that can be realized using a publish/subscribe communication infrastructure implicitly rely on ordering guarantees. In the following, concise descriptions of different options are given which have also been identified in research on group communication. It is important to note that ordering guarantees always come at a cost since they limit concurrency in the system.

Total Ordering. For some classes of applications it is a basic requirement that all subscribers receive every notification in the *same* order. In combination with message completeness, total ordering resembles the "atomic broadcast primitive" introduced by Birman and Joseph [23], where total ordering is combined with the requirement that either *all* subscribers receive the message or *none*.

Total ordering can be used, for example, to keep replicated copies of databases consistent. In this case, all commands that alter the contents of the database are sent to all copies in parallel. In this scenario, it is important that all database copies receive the database commands in the same order. Otherwise, the contents of different copies may diverge, for example, if two update commands for the same data item are received in different ordering at two copies.

It is important to note that total ordering does not require any particular order at individual receivers, i.e., total ordering can be combined with the orderings described below. In a publish/subscribe system, it is very difficult to realize total ordering because of the loose coupling and the anonymous communication.

FIFO-Publisher Ordering. Imagine a set of actuators which can open or close valves according to a command they receive encapsulated in a notification from one publisher: if the valve is open and the commands "open" and "close" are received in this order, the state of the valves will be closed afterwards. However, if the ordering is mixed up, the final state of some valves may be "open" with possibly serious consequences. In this case, *FIFO-publisher ordering* is needed. However, it is not necessary, that all subscribers receive all notifications in the same order in this case, because notifications from another publisher that trigger other functionalities at the actuators can be arbitrarily interleaved with the valve-commands depending on the application.

Causal Ordering. Another ordering requirement is *causal ordering*. Here, notifications arrive in FIFO-publisher ordering but cannot be arbitrarily interleaved with notifications received from other publishers. Instead, it is guaranteed that a message m_2 which is triggered by another message m_1 arrives after m_1 at every node—even if the publishers of m_1 and m_2 are different. Causal ordering has been introduced together with the *happened-before* relation by Lamport [99] and was later applied to multicast communication by Birman and Joseph as part of the *causal broadcast primitive* [23].

Bounded Delay

Reconfiguring of the broker topology, especially in combination with ordering guarantees, may incur an overhead which can result in delayed message delivery. In real-time systems, for example, it might be required that the maximum delay is bounded. The same applies to the time it takes to implement the reconfiguration. Some applications may require an upper bound for the time needed for asynchronous communication and possible coordination.

The time it takes for a reconfiguration to be implemented correlates with link delays and processing capabilities of the brokers involved as well as with the type of the reconfiguration. This guarantee is not considered here and is left open for future work.

4.4 Reconfiguring Broker Overlay Topologies

In this section, an algorithm for reconfiguring regular acyclic broker topologies of publish/subscribe systems is presented, which prevents message loss and guarantees ordering in face of reconfigurations. Thereby, the focus is on elementary reconfigurations and the difficult case of exchanging links in particular, because adding and removing leaf brokers is easy as discussed above.

The algorithm builds on the important observation which has first been published by Cugola et al. in [50]: when replacing a link, only those brokers' routing tables are affected that are part of the reconfiguration cycle. The rationale behind this insight is that exchanging a link does not inject any new subscription or advertisement into the system. Thus, the set of notifications or (un)subscriptions forwarded to or from the brokers on the reconfiguration cycle does not change. This can easily be seen by "contracting" all the brokers on the reconfiguration cycle to one logical broker. Doing this, the set of notifications forwarded to or by this logical broker must not change due to reconfigurations inside this logical broker. The approach taken here is, thus, to limit all reconfiguration activities to those brokers on the reconfiguration cycle. This way, it is possible to minimize the overhead that is introduced by reconfigurations. The basic idea is to remove all subscriptions from the routing tables of the brokers on the reconfiguration cycle that are needed to route notifications over the old link and instead install routing entries for the new link.

The following discussion of related work starts by explaining the most basic related approaches which rely only on standard publish/subscribe operations and do not need any coordination efforts. Based on this, it is motivated that coordination is an important issue during reconfiguration. Related approaches are discussed which try to overcome the problems inherent to the uncoordinated approaches by additional coordination. Then, a new approach is presented which is called the *advanced coordinated reconfiguration algorithm.* Finally, an evaluation of the new reconfiguration algorithm based on a simulation study is presented.

4.4.1 Related Work

This section starts with the simplest approach to reconfiguration which employs no coordination between brokers. Then, more sophisticated approaches that introduce coordination mechanisms and remove several shortcomings are discussed.

Uncoordinated Approaches

The easiest way to activate a new link a and deactivate the old link r in the routing configuration is to issue unsubscriptions regarding r and issuing respective subscriptions regarding a. The approaches that follow this recipe are called *uncoordinated* in the following if the brokers on r do not coordinate their actions with the brokers at a. They are commonly used in practice and are in general easy to integrate into publish/subscribe systems since they can be layered on top of the routing layer and require no further modifications on the lower layers.

Strawman Approach. The first approach to reconfiguring an acyclic publish/subscribe broker overlay topology has been published by Yu et al. [162]. It follows a straightforward approach: when link r is replaced by another link a, the two brokers B_1^r and B_2^r connected by r act as if they received unsubscriptions for all subscriptions they received in the past from the respective other side of r. Analogous, the two brokers B_1^a and B_2^a connected by a forward all the subscriptions they received in the past from their neighbor brokers to the respective broker on the other side of a, where they are handled as regular subscriptions and disseminated accordingly. As a result, notifications which have been routed over r will be routed over a afterwards. In accordance with the literature, this approach is called the STRAWMAN approach in the following.

The two most important advantages of the STRAWMAN approach are that (i) reconfigurations can easily be carried out in parallel and that (ii) they can be implemented by simply using the standard publish/subscribe operations sub() and unsub(). However, it has also three serious drawbacks which consist of (i) a potentially large message overhead, (ii) possible message loss, and (iii) possible message reordering. The potentially large overhead results from unsubscriptions issued by brokers connected to r that leave the reconfiguration cycle and are later reestablished by subscriptions issued by the brokers connected to a (i.e., unsubscriptions, and respective subscriptions, may leave the reconfiguration cycle). Messages may be lost due to race conditions when the brokers connected to r and a act in an uncoordinated way as described in Section 4.3.1.

Later, the STRAWMAN approach is used as a baseline for the evaluation of the presented algorithm because it is a well-known representative of the uncoordinated approach and often used for comparison in literature.

Deferred Unsubscriptions. To reduce the number of unsubscriptions that can potentially leave the reconfiguration cycle, the DEFERRED UNSUBSCRIPTIONS algorithm has been proposed which slightly improves the STRAWMAN approach [64]. The contribution is simply adding a delay between the time the brokers on a start issuing their subscriptions and the time the brokers connected to r issue their unsubscriptions. In the ideal case, this delay should correspond to the time that is needed to set up the new link and propagate the subscriptions of the brokers connected to it. However, this approach still bears a significant disadvantage. If a bigger value is chosen for the timeout, the probability of message loss and unsubscriptions leaving the reconfiguration cycle decreases while the probability of duplicates increases simultaneously, since notifications may be routed over r *and* a. The opposite is true for smaller values for the timeout. It is also in general not possible to avoid unnecessary overhead, message loss, or message duplication this way—especially in the face of nondeterministically varying communication and processing delays.

Coordinated Approaches

From the problems of the uncoordinated approaches described above it becomes apparent that coordination between the brokers on the reconfiguration cycle can prevent message loss and limit the overhead. In the following, related approaches are described which invest in coordination to different degrees in order to minimize overhead, message loss, and duplicates.

Coordinated Unsubscriptions. The authors of [52] propose the COORDINATED UNSUBSCRIPTIONS algorithm which uses dedicated FLUSH messages in addition to a timeout as above in order to ensure that the endpoints of a have issued their subscriptions before the endpoints of r start sending out their respective unsubscriptions. By doing this, it is ensured for many routing algorithms that no unsubscriptions leave the reconfiguration cycle. The FLUSH message is sent by the brokers on a right after they issued their subscriptions and marks the end of subscription forwarding. When brokers at r receive this message, they can start issuing their unsubscriptions. The FLUSH message is meant to be flooded through the whole tree which may result in a significant overhead.

The scenario the authors target consists of highly dynamic environments, where links are removed spontaneously. Here, they achieve much better results than the STRAWMAN solution since unsubscriptions normally do not leave the reconfiguration cycle [126]. However, this scenario is different from the scenario of reconfigurations targeted here, where links are exchanged in a controlled manner. If r is removed instantaneously, this may result in message loss with coordinated unsubscriptions, while waiting with tearing down r until both endpoints of a have issued their subscriptions may lead to duplicates (as discussed above for the DEFERRED UNSUBSCRIPTIONS algorithm). Another issue not addressed by the authors but important in many scenarios is that of notification ordering.

Informed Link Activation. The INFORMED LINK ACTIVATION algorithm which improves the DEFERRED UNSUBSCRIPTIONS algorithm above is presented by Frey in [64]. The aim of the author is to further reduce the number of subscriptions propagated during reconfiguration. Therefore, the endpoints of r determine the set of subscriptions for which only brokers are subscribed that belong to the tree on the "other side" of r. These subscriptions are then forwarded to the brokers connected to the respective side of a which accordingly handle them as subscriptions they received from a neighbor broker on "their side" of a. Similarly, the brokers on r only unsubscribe for subscriptions from this set. Thereby, the number of (un)subscriptions issued during reconfiguration is reduced because new unsubscriptions forwarded on the reconfiguration cycle that have not reached r at the beginning of the reconfiguration would otherwise lead to subscriptions forwarded over a which would later be removed when the unsubscriptions issued reach the brokers at a. For the actual coordination of (un)subscriptions, the author proposes the use of one of the algorithms discussed above.

Although this algorithm reduces the overhead during reconfiguration it also does not prevent message loss or give any ordering guarantees because it still relies on the algorithms discussed above for coordination purposes.

Reconfiguration Path. A more sophisticated approach is taken in [50], where the RECONFIGURATION PATH algorithm is proposed which uses additional messages, similar to the informed link activation algorithm above, to limit the overhead of reconfigurations even more. It invests more effort into keeping all reconfiguration related actions limited to the reconfiguration cycle (or *reconfiguration path* how the authors call it). Therefore, they introduce dedicated messages which do not leave the reconfiguration path such that the (un)subscriptions that are necessary to adapt the routing tables of the brokers on the reconfiguration path do never leave it. This way, it is possible to limit the (un)subscriptions sent on the cost of a special treatment of concurrent (un)subscriptions which are forwarded as part of the normal operation of the publish/subscribe system. The advantage gained is that no messages are flooded anymore through the whole broker network. Instead, the brokers at r send a reconfiguration message over the reconfiguration cycle which includes the subscriptions to be removed for r and the subscriptions to be added for a regarding their side of the reconfiguration cycle. When the reconfiguration message has reached the broker on the other side of r then the routing tables on the reconfiguration cycle are prepared for adding a from the perspective of the original issuer of the reconfiguration message.

Although very sophisticated, this approach also lacks the requirement of keeping notifications ordered. Additionally, notifications still might get lost during the reconfiguration process when the reconfiguration message is being forwarded. This is not a serious disadvantage in some scenarios, where notification loss cannot be avoided at all because links may vanish unexpectedly. Here, where reconfigurations should not disrupt the system service, however, this algorithm is not applicable. Another disadvantage of the RECONFIGURATION PATH algorithm is that parallel reconfigurations are not supported if the reconfiguration cycles overlap because brokers can only attend one reconfiguration at a time.

Summary

From the approaches discussed above it becomes evident that coordination can be of substantial help to be able to provide the guarantees identified in Section 4.3.3. All the algorithms presented so far were not able to prevent message loss which is an important requirement for non-disruptive reconfigurations of publish/subscribe broker overlay topologies.

It also became clear that message ordering guarantees can be achieved by delaying them appropriately. Thus, a reconfiguration is implemented by introducing additional queues and messages which coordinate the brokers involved in the reconfiguration process.

4.4.2 Advanced Coordinated Reconfiguration

In this section, a new algorithm is described for exchanging links to seamlessly reconfigure the broker overlay topology of a publish/subscribe system. It starts with introducing a coloring mechanism which is applied to prevent message loss and keep message orderings. Then, rules are formulated which have to be respected to prevent message loss and guarantee message ordering. With those rules it is easier to follow the explanations of the algorithm.

In the following, the labeling introduced in Figure 4.3 on page 105 will be used. The link to be replaced is called r, the new link a, and the brokers connected to them $B_{1/2}^r$ and $B_{1/2}^a$, respectively. Furthermore, the algorithms will be illustrated where sensible only for the left side of the reconfiguration cycle, i.e., the side to which B_1^r and B_1^a belong, as actions for the right side are analogous.

The basic idea of the algorithm is that all brokers are colored black initially and that the brokers connected to r send the subscriptions regarding the other side of the reconfiguration cycle out-of-band to the brokers on the respective side of a. Those brokers handle them as if they received them from the other side of a and forward them accordingly. Then, the brokers at a send a special message in direction of the brokers at r which colors the brokers on the path gray and on which receipt the brokers at r issue unsubscriptions regarding r. To be able to provide ordering guarantees, messages may need to be queued at the brokers connected to a. This affects notifications and (un)subscriptions that are forwarded during the reconfiguration process.

The algorithm requires simple, identity-based, or covering-based routing on the notification routing layer. The reason for this is that it is assumed that it suffices to send the routing table entries of the broker on one side of r to the broker on the same side of a in order to be able to reach a consistent state of the routing tables of the brokers on the reconfiguration cycle. This assumption only holds if the routing entries in the routing tables of the brokers at a are the same after the subscriptions forwarding is completed as if the (un)subscriptions in the past had been issued in the new topology which contains a but not r. This is true for the aforementioned routing algorithms but does not necessarily hold for every routing algorithm. Furthermore, advertisements are initially not supported but it will be discussed later how they can be incorporated.

Coloring

To ease reasoning, a coloring mechanism is adopted in the following. When starting a reconfiguration, every broker that is affected by it (i.e., every broker on the reconfiguration cycle) is colored *black*. A broker that has received all the subscriptions that enable him to route notifications to the other side of the new link turns its color to *gray*.

As discussed above, the following approach is taken: $B_{1/2}^r$ determine all the subscriptions that are needed to route notifications over r and send them to their counterparts at a on the same side of the reconfiguration cycle. $B_{1/2}^a$ add the subscriptions to their routing tables such that all notifications routed over r can now be routed over a. They disseminate the subscriptions received on their side of the reconfiguration cycle just like they received the subscriptions from the other side of a. Doing this, the brokers on the path from B_1^a (B_2^a) to B_1^r (B_2^r) gradually turn their color from black to gray since they—and every other broker on the path to a—are now capable of forwarding notifications over a.

Every notification that enters the reconfiguration cycle inherits the color of the first broker it encounters on the reconfiguration cycle. The same applies to (un)subscriptions. This information is later used to decide whether a message can be forwarded or must be delayed.

Guarantees

With the coloring mechanism and the basic forwarding strategy, it is now possible to formulate rules which must hold to not break the guarantees which are to be provided here. They are derived from the guarantees identified in Section 4.3.3 on page 114.

Duplicates and Message Completeness. To be able to guarantee the prevention of message duplicates or message loss, the following rules must hold.

Rule 1 (Notification Duplicates). *During the reconfiguration process there are two* *links, r and a, which connect the left and the right subtree. To avoid duplicates, notifications must either be sent over r or over a.*

Each notification has one of the two colors according to the coloring mechanism because every notification inherits the color of the first broker on the reconfiguration cycle it encounters. Rule 1 is implemented by routing black notifications only over r and gray notifications only over a.

Rule 2 (Notification Loss). *The old link r must not be removed if there is still a notification on the reconfiguration cycle that will be routed over it.*

Again, it is possible to take advantage of the coloring mechanism here and implement Rule 2 by not removing r until all brokers on the reconfiguration cycle have turned gray. The following lemmas are needed to show that this is sufficient to satisfy Rule 2.

Lemma 3. *Removing r before all brokers have turned gray may lead to lost notifications.*

Proof. If the old link r would be removed before all brokers have turned gray, this may lead to the following situation: a notification arrives at a black broker and, thus, turns black. Accordingly, the message cannot be forwarded over a because of its color. Since r has already been removed, it cannot be forwarded over r, and because the message is black, it cannot be forwarded over a. Thus, the message is lost. □

Lemma 4. *When all brokers on the reconfiguration cycle have turned their color to gray, there are no more black messages on the reconfiguration cycle that need to be routed over r.*

Proof. When all brokers have turned gray, every new notification that enters the reconfiguration cycle will turn gray, too. The order in which the brokers turn gray starts from the brokers connected to a and ends at the brokers connected to r. Since the recoloring of the brokers is stimulated by a message that is forwarded from $B^a_{1/2}$ to $B^r_{1/2}$ and FIFO channels are required between the brokers, every black message that is forwarded on the reconfiguration cycle after all brokers have turned gray cannot stem from this side of the reconfiguration cycle and, thus, does not need to be forwarded over r anymore because it has already passed it then.   □

For (un)subscriptions, the same rules apply that were discussed above for notifications. Due to the coloring mechanism that applies to notifications as well as (un)subscriptions, the implementation of Rule 2 is analogous to the implementation of Rule 3 below.

Rule 3 ((Un)subscription Loss). *To not lose any (un)subscription, the old link r must not be removed if there is still a (un)subscription that has to be routed over r.*

The same holds for (un)subscription duplicates.

Rule 4 ((Un)subscription Duplicates). *To avoid (un)subscription duplicates, they must only be routed either over r or over a.*

In order to reduce message overhead, this rule is implemented by simply forwarding all (un)subscriptions that are received over r to the respective broker at a. A broker connected to a that receives such a (un)subscription handles it as if it had received it over a. This mechanism will be discussed in greater detail later.

(Un)subscription Ordering. After handling message completeness and duplicates, message ordering is discussed now. First, the ordering of (un)subscriptions, which has to be taken care of separately, is discussed: if a broker issues a subscription followed by an unsubscription it is necessary that both messages arrive in the same order they were issued. Thus, every broker must receive subscription/unsubscription pairs issued by the same broker in exactly the same order they were published. This can be generalized to the requirement that all (un)subscriptions must be received in FIFO-producer ordering.

Ensuring FIFO-producer ordering for (un)subscriptions implies that they may need to be delayed. An example for a scenario, where this is necessary is illustrated in Figure 4.6 (for ease of understanding, it shows only the messages forwarded on the reconfiguration cycle). Here, broker B_3 issues a subscription s which enters the reconfiguration cycle at broker B_1 which is colored black at this time. Shortly after B_1 forwarded s from Broker B_3 it turns gray such that the following unsubscription $\neg s$ issued by B_3 turns gray when it is forwarded by B_1. Due to the different colors, s is forwarded over r to reach the other side of the reconfiguration cycle, while $\neg s$ is forwarded over a. Longer delays on the path $\overline{B_1 B_1^r B_2^r B_2 B_2^a}$ than on the path $\overline{B_1 B_1^a B_2^a}$ may have the effect that B_2^a receives $\neg s$ before s and, hence, has a stale routing entry for s afterwards. Thus, Rule 5 can be deduced.

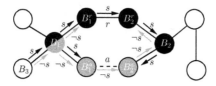

Figure 4.6: Example scenario, where B_2^a may not receive (un)subscriptions in FIFO-producer ordering

Rule 5 ((Un)subscription FIFO-Producer Ordering). *(Un)subscriptions must be received in FIFO-producer ordering by every broker in the system. To avoid that (un)subscriptions overtake each other, it is necessary to delay gray (un)subscriptions which stem from this side of the reconfiguration cycle from being sent over a until it is assured that all black (un)subscription that stem from this side left the other side of the reconfiguration cycle.*

To be able to delay (un)subscriptions in order to implement Rule 5 queues are introduced. Thereby, it is possible to maintain their FIFO-producer ordering.

To minimize the message complexity of (un)subscription forwarding during reconfiguration, the following optimization are proposed. Assume that broker B_3 in

Figure 4.7(a) issues a subscription s when the reconfiguration has already started. As depicted, it arrives at B_1 and inherits the color black. Accordingly it is forwarded to B_1^r and B_1^a. Since its color is black, it will also be forwarded over r to B_2^r but not over a. This subscription will later be removed by B_2^r after being installed by B_2^a. Thus, regular processing of (un)subscription forwarding is changed at this point in order to reduce the message overhead. When s arrives at B_2^r it is not forwarded on the right side but instead colored gray and sent to B_2^a, where it is handled as if B_2^a had received it from B_1^a. This is done, because it is not sensible to establish the routing entries on the right side of the reconfiguration cycle for s starting from B_2^r when r will be removed and s has to be migrated to B_2^a later anyway.

When B_3 issues a respective unsubscription shortly after s, but still long enough such that B_1 turned its color into gray meanwhile, the situation depicted in Figure 4.7(b) appears. Similar to s, the unsubscription $\neg s$ is routed over B_1 to B_1^r and B_1^a. Since its color is gray, it is not routed over r but over a instead. In this situation, it is important that s is processed before $\neg s$ at B_2^a. Otherwise it could happen that $\neg s$ is disseminated before s in the right part of the tree which would lead to a situation, where the subscription s exists in the routing tables of brokers on the right side but there is no subscriber on the left side. Therefore, a queue is added to B_2^a to delay gray (un)subscriptions (including $\neg s$) which have not been relayed from a broker on the same side of the reconfiguration cycle until all black (un)subscriptions from the other side have left this side of the reconfiguration cycle.

The queue for (un)subscriptions is called Q_{sub} in Figure 4.7. When B_2^a receives a gray (un)subscription from B_2^r it disseminates it like an (un)subscription it received from B_1^a. If the (un)subscription is gray and received over a, it is put into Q_{sub}. (Un)subscriptions in Q_{sub} are handled during the reconfiguration, when it is guaranteed that no black (un)subscriptions will arrive at B_2^a which stem from the other side of the reconfiguration cycle, since this would disturb the required FIFO-producer ordering.

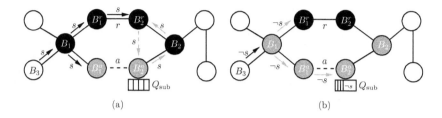

Figure 4.7: Realizing FIFO-producer ordering of (un)subscriptions using a queue

Notification Ordering. The same mechanism introduced can be used to implement (un)subscription ordering in Rule 5 to guarantee FIFO-publisher ordering for notifications because the situation is similar. Figure 4.8 depicts a situation, where B_4 subscribed to a filter that matches notifications published by B_3 (the small arrows represent the respective routing table entries). Now assume that B_3 publishes n_1 which enters the reconfiguration cycle at B_1 when this broker is still black (Figure 4.8(a)). Accordingly, n_1 is colored black and routed over r. If B_3 now publishes n_2 and n_2 enters the reconfiguration cycle right after B_1 turned gray, it is colored gray and not routed over r. Instead, it is routed over a, where it is queued analogous to gray (un)subscriptions at B_2^a in the queue Q_{not}. This prevents the case that n_2 arrives at B_4 before n_1 (Figure 4.8(b)). The problem is analogous to that of (un)subscription FIFO-producer ordering with the exception that n_1 is forwarded regularly by B_2^r.

For queueing notifications and (un)subscriptions, separate queues are used since (un)subscriptions must always be delivered in FIFO-producer ordering while notifications may be delivered without any ordering guarantees or with a separate ordering, depending on the application requirements.

The following rule applies to notification FIFO-publisher ordering analogous to Rule 5.

Rule 6 (Notification Queueing for FIFO-Publisher Ordering). *Gray notifications that are sent over a must be queued and delayed at a broker $B_{1/2}^a$ until it is guaranteed that all black notifications which stem from the other side of the reconfiguration*

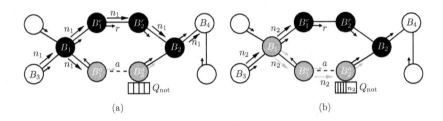

(a)   (b)

Figure 4.8: Realizing FIFO-publisher ordering of notifications with a queue

cycle have left the half of the reconfiguration cycle $B_{1/2}^a$ belongs to.

The reasoning is similar to that of Rule 5: if there is still a black notification on the right side of the reconfiguration cycle which stems from the left side of the reconfiguration cycle, it is possible that a dequeued notification reaches a broker on B_2^a's side of the broker tree before the black notification arrives there. If both notifications stem from the same publisher, the FIFO-publisher ordering is disturbed in this case.

In Section 4.3.3, causal ordering has been identified as another important ordering besides FIFO-publisher ordering. It requires that a notification n_2 which depends causally on another notification n_0 must be received after n_0 at every subscriber. This problem is particularly difficult during reconfiguration if n_0 is published on a broker which belongs to the subtree on one side of the reconfiguration cycle while n_2 is published on the subtree which belongs to the other side.

For example, assume that B_4 in Figure 4.8(a) publishes n_0 (which is colored black accordingly) for which B_3 and B_2^a are subscribed while B_3 publishes n_1. Now, assume that B_3 publishes n_2 in reaction to receiving n_0 (as depicted in Figure 4.8(b)) and that B_1 turned gray, such that n_2 is colored gray, too. Furthermore assume, that B_2^a is also subscribed for a filter matching n_2. Then, it is required that a subscriber connected to B_2^a which is subscribed for filters matching all three notifications receives n_2 *after* n_0 and n_1.

In the example, it is necessary that the delivery of n_2 must be delayed until n_1 and n_0 have reached B_2^a. Since it could take a long time to forward n_1 on the reconfiguration cycle, the generalized rule that n_2 in Q_{not} at $B_{1/2}^a$ must be delayed until there is no more black notification on the whole reconfiguration cycle can be derived. The following rule summarizes this.

Rule 7 (Notification Queueing for Causal Ordering). *Gray notifications that are queued in Q_{not} at broker $B_{1/2}^a$ have to be delayed until it is guaranteed that no more black notifications will be routed over the reconfiguration cycle until the end of the reconfiguration.*

It is important to note that queuing notifications in Q_{not} is only necessary if ordering guarantees are required. Otherwise, notifications arriving at B_2^a can be forwarded instantaneously.

Integration Into the Publish/Subscribe Model

The advanced coordinated reconfiguration algorithm relies on queues and extra messages which are needed by the brokers to ensure certain guarantees, for example, that all black messages have left the reconfiguration cycle in the case of causal ordering. Thus, it is necessary to include these additions into the publish/subscribe system model. Figure 4.9 depicts the model of a broker B together with the queues Q_{not} and Q_{sub} needed for the reconfiguration (as mentioned above, Q_{not} can be omitted if a certain notification ordering is not required). When B receives a message it enqueues it in the respective queue if necessary. Therefore, the receiveMsg() procedure has to be modified such that incoming messages can be enqueued before they are processed by the handleMsg() procedure and are colored appropriately. The procedure sendMsg() must also be altered such that the color attribute is resetted when a message leaves the reconfiguration cycle and such that messages are only sent if they are colored appropriately.

In total, five new message types are introduced for the integration of the new reconfiguration algorithm into the publish/subscribe model:

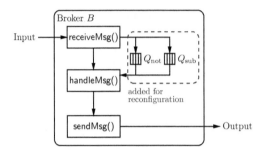

Figure 4.9: Integration of queues needed for the advanced coordinated reconfiguration algorithm into the publish/subscribe broker model

Lock: With the *lock message*, the initiator of the reconfiguration process "locks" all brokers on the reconfiguration cycle such that they cannot be part of another reconfiguration. This is done to avoid disturbances due to parallel overlapping reconfigurations which are not supported by the algorithm and may lead to undesired side effects.

Begin: The initiator of the reconfiguration process sends a *begin message* to the brokers $B^r_{1/2}$ connected to r. Subsequently, $B^r_{1/2}$ determine the set of subscriptions P_{sub} they received from the broker on the other side of r and send them out-of-band to the respective broker $B^a_{1/2}$ on the same side of the reconfiguration cycle connected to a. $B^a_{1/2}$ then handle these subscriptions as if they received them over a and disseminate them accordingly.

Separator: The two initial *separator messages* are sent by $B^r_{1/2}$ to $B^a_{1/2}$ following the subscriptions regarding r. The receiving brokers forward it along the path to $B^a_{1/2}$ on the reconfiguration cycle when they have established all routing entries for a that they received previously. Forwarding the separator message implies that a broker has turned its color to gray which also means that it will not inject any black message into the reconfiguration cycle anymore because every message that reaches the reconfiguration cycle through this broker will be colored gray. When the **separator** message finally arrives at the brokers on the respective side of r, $B_{1/2}^r$ can be sure that all brokers on their side have turned gray.

In order to inform all brokers on the reconfiguration cycle about this, they forward the **separator** message over r towards $B_{2/1}^r$ on the other side of the reconfiguration cycle and back to $B_{1/2}^a$ on their respective side of the reconfiguration cycle after they issued unsubscriptions regarding r. If $B_{1/2}^r$ received both **separator** messages it starts unlocking the reconfiguration cycle since the reconfiguration is then finished from its perspective as all brokers on the reconfiguration cycle have turned gray then and the unsubscriptions regarding r have already been issued. If $B_{1/2}^a$ receive a **separator** message from the other side, they can start dequeuing notifications in Q_{not} depending on the ordering requirements.

End: When the brokers connected to r will relay no more subscriptions regarding the old link r (which is the case when they have received a **separator** message over r), they send an *end message* to the respective brokers connected to the same side of a. The receiving broker $B_{1/2}^a$ then knows that the sending broker has received and processed the **separator** message and that all brokers on the other side have turned gray. Accordingly, it will not receive any forwarded subscriptions anymore from the sender. Thus, it can now start to dequeue the subscriptions in Q_{sub}.

Unlock: When the reconfiguration is ended, the reconfiguration cycle is unlocked with an *unlock message*. This means that brokers on the reconfiguration cycle are now free to join another reconfiguration. The initial unlock message messages are issued by the brokers at r on the receipt of a **separator** message.

Algorithm Details

In this section, it is described in detail how the advanced coordinated reconfiguration algorithm works. The description steps chronologically through the algorithm by concisely explaining the five message types, when, how, and why they are sent. The algorithms are formulated such that they fit into the routing framework proposed in [115].

Variables. The algorithms use several variables. The broker which runs the algorithm refers to itself using B_{this} and its neighbor brokers are stored in the set $\mathcal{B}_{\text{neighbors}}$. Every broker on the reconfiguration cycle maintains two pointers B_{left} and B_{right} that point to the preceding and succeeding broker on the cycle (from the perspective of the initiator), respectively.

Similarly, on brokers connected to r (a), B_{relay} stores a pointer to the broker on the other side of r (a). The brokers connected to r (a) use B_{new} (B_{old}) to store a pointer to the respective broker on the other side of the link.

The set $\mathcal{B}_{\text{gray}}$ stores pointers to the brokers that already turned from black to gray. This information is later used to determine if there are still black messages on one or both sides of the reconfiguration cycle. The brokers connected to a (i.e., B_1 and B_n) use $\mathcal{B}_{\text{queue}}$ to store the set of brokers from which they have not yet received a **separator** message. This set is only needed for notification ordering.

Table 4.1 gives an overview of the variables with a short description. Figure 4.10 depicts the pointers in the example scenario. In the following, sometimes B_1^a is used instead of B_1 and B_2^a instead of B_n where it improves the readability.

Lock Message. The reconfiguration process is started by one broker (B_1), called the *initiator*, by executing the handleLockMsg() procedure described in Algorithm 2. The initiator is one end-point of the new link a $(\overline{B_1 B_n})$. It is responsible of implementing the reconfiguration and is initially the only broker which knows about the reconfiguration cycle, a, and r. It starts the reconfiguration by sending out a lock message in order to lock the reconfiguration cycle. A reconfiguration is only executed, if every broker on the reconfiguration cycle has been locked successfully, avoiding race conditions due to parallel overlapping reconfiguration processes thereby. For routing purposes, the lock message m^{l} contains the

Variable	Description
B_{this}	The broker running the algorithm.
B_{left}	Preceding broker on the reconfiguration cycle.
B_{right}	Succeeding broker on the reconfiguration cycle.
$\mathcal{B}_{\text{neighbors}}$	Set of neighbor brokers of B_{this}.
B_{old}	Points to the broker on the other side of r (only B_1^r and B_2^r).
B_{new}	Points to the broker on the other side of a (only B_1 and B_n).
B_{relay}	Points to the broker on the same side of r (a) if the broker is connected to a (r) (only B_1, B_1^r, B_2^r, and B_n).
$\mathcal{B}_{\text{gray}}$	Set of brokers that turned their color from black to gray.
$\mathcal{B}_{\text{queue}}$	Set of brokers no **separator** message has been received from yet (only B_1 and B_n).
relaying	States if the brokers on r relay (un)subscriptions to B_1 and B_n, respectively (only B_1^r and B_2^r).

Table 4.1: Overview of variables used during reconfiguration

brokers on the reconfiguration cycle from the initiator to the other endpoint of a ($m_{\text{path}}^l = \langle B_1, \dots, B_1^r, B_2^r, \dots, B_n \rangle$) plus the edge to be removed ($m_{\text{edge}}^l = \langle B_1^r, B_2^r \rangle$).

The lock message brings the receiving broker B into the state "locked" and changes its color to black. Due to parallel reconfigurations, it may happen that B is already locked by another reconfiguration process when it receives m^l (line 39). In this case, and if the topology on the reconfiguration cycle changed due to other reconfigurations such that the neighbor brokers on the path m_{path}^l do not share a link anymore (line 23), B sends m^l back on the path to the initiator (in case B is not the initiator itself). The receiving brokers accordingly change their state back to "unlocked" until the initiator received the message (line 33).

The lock request succeeded, i.e., all brokers on the reconfiguration cycle have been locked successfully, if the initiator of the lock request (the first broker stored in m_{path}^l) receives the lock message again. In this case, B_1 adds B_n to its neighbors and sends out a begin message m^b to the endpoints of r (B_1^r and B_2^r) in order to start the actual reconfiguration (line 28). After the handleLockMsg() procedure has terminated at a broker, it is locked, colored black but forwards messages and (un)subscriptions as before.

Figure 4.10 depicts an example scenario, where B_2 has just processed a lock message received from B_1. Accordingly, it has set its pointer variables to the values shown. Similarly, B_1 has also set its pointers as depicted.

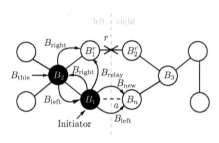

Figure 4.10: Example scenario depicting variable values after B_2 processed the lock message m^l

Begin Message. After the brokers on the reconfiguration cycle have been locked successfully, the initiator starts the reconfiguration by sending a begin message m^b to the brokers B_1^r and B_2^r connected by r. The receiving brokers accordingly collect the subscriptions which point to the respective broker $B_{2/1}^r$ on the other side of r (B_{relay}) and send them to the broker $B_{1/2}^a$ out-of-band on their side of the reconfiguration cycle (B_{relay}) in form of a subscription message m^r (Algorithm 3, line 8). These subscriptions are marked as relayed and are colored gray.

After all subscriptions for the other side have been forwarded, $B_{1/2}^r$ send a separator message m^s to their respective B_{relay} (line 12). This marks the end of the subscription forwarding phase.

In the case that B_{old} from the perspective of $B_{1/2}^r$ has already turned gray (i.e., the separator message from B_{old} has already reached B_{old} again, and also $B_{1/2}^r$ to disseminate this information), $B_{1/2}^r$ sends an end message m^e right after the separator message to B_{relay} (line 15). This can happen due to message forwarding delays which may differ significantly on both sides of the reconfiguration cycle.

Algorithm 2 Handle lock message m^l

$m^l_{\text{path}} = \langle B_1, \ldots, B^r_1, B^r_2, \ldots, B_n \rangle$, $m^l_{\text{edge}} = \langle B^r_1, B^r_2 \rangle$

1   Set **procedure** handleLockMsg(Broker B_{sender}, LockMsg m^l)
2   **begin**
3   **if** $\neg$locked **then** // *initialize variables*
4      $B_{\text{left}} \leftarrow B_{\text{this}} \neq B_1$? B_{i-1} : B_n
5      $B_{\text{right}} \leftarrow B_{\text{this}} \neq B_n$? B_{i+1} : B_1
6      $\mathcal{B}_{\text{neighbors}} \leftarrow B_{\text{right}} = B_1$? $\mathcal{B}_{\text{neighbors}} \cup B_1$: $\mathcal{B}_{\text{neighbors}}$
7   **if** $B_{\text{right}} \in \mathcal{B}_{\text{neighbors}}$ **then** // *accept lock message*
8        $B_{\text{new}} \leftarrow \perp$; $B_{\text{old}} \leftarrow \perp$; $B_{\text{relay}} \leftarrow \perp$; $\mathcal{B}_{\text{gray}} \leftarrow \perp$
9   **if** $B_{\text{this}} = B_1$ **then** $B_{\text{relay}} \leftarrow B^r_1$; $B_{\text{new}} \leftarrow B_n$
10   **elseif** $B_{\text{this}} = B_n$ **then** $B_{\text{relay}} \leftarrow B^r_2$; $B_{\text{new}} \leftarrow B_1$
11   **elseif** $B_{\text{this}} = B^r_1$ **then** $B_{\text{relay}} \leftarrow B_1$; $B_{\text{old}} \leftarrow B^r_2$
12   **elseif** $B_{\text{this}} = B^r_2$ **then** $B_{\text{relay}} \leftarrow B_n$; $B_{\text{old}} \leftarrow B^r_1$
13   **endif**
14        $\mathcal{B}_{\text{queue}} \leftarrow \emptyset$
15   **if** $B_{\text{this}} \in \{B_1, B_n\} \wedge$ FIFO order **then**
16          $\mathcal{B}_{\text{queue}} \leftarrow B_{\text{this}} = B_1$? $\{B^r_2, \ldots, B_n\}$: $\{B_1, \ldots, B^r_1\}$
17   **elseif** $B_{\text{this}} \in \{B_1, B_n\} \wedge$ causal order **then**
18          $\mathcal{B}_{\text{queue}} \leftarrow \{B_1, \ldots, B_n\}$
19   **endif**
20   side $\leftarrow B_{\text{this}} \in \{B_1, \ldots, B^r_1\}$? left : right
21   color $\leftarrow$ black ; locked $\leftarrow$ true ; relaying $\leftarrow$ false
22        $\mathcal{M} \leftarrow \{(B_{\text{right}}, m^l)\}$
23   **else** // *path changed meanwhile*
24        $B_{\text{left}} \leftarrow \perp$; $B_{\text{right}} \leftarrow \perp$
25        $\mathcal{M} \leftarrow \{(B_{\text{sender}}, m^l)\}$
26   **endif**
27   **else** // *broker is already locked*
28   **if** $B_{\text{sender}} = B_{\text{left}}$ **then** // *lock request succeeded*
29        $\mathcal{B}_{\text{neighbors}} \leftarrow \mathcal{B}_{\text{neighbors}} \cup B_{\text{left}}$
30        $m^b \leftarrow$ new BeginMsg() // *start reconfiguration with a* **begin** *message*
31   sendMsg(B^r_1, m^b) ; sendMsg(B^r_2, m^b)
32        $\mathcal{M} \leftarrow \emptyset$
33   **else if** $B_{\text{sender}} = B_{\text{right}}$ **then** // *lock request failed*
34        $\mathcal{M} \leftarrow B_{\text{new}} = \perp$? $\{(B_{\text{left}}, m^l)\}$: $\emptyset$
35        $B_{\text{left}} \leftarrow \perp$; $B_{\text{right}} \leftarrow \perp$; $B_{\text{old}} \leftarrow \perp$; $B_{\text{relay}} \leftarrow \perp$
36        $\mathcal{B}_{\text{queue}} \leftarrow \emptyset$
37   locked $\leftarrow$ false
38   **else** // *concurrent reconfiguration*
39        $\mathcal{M} \leftarrow B_{\text{sender}} \neq B_{\text{this}}$? $\{(B_{\text{sender}}, m^l)\}$: $\emptyset$
40   **endif**
41   **endif**
42   **return** $\mathcal{M}$
43 **end**

The handleBeginMsg() procedure returns an empty set because all the messages to be sent out have already been sent in the body of the procedure (line 19). Additionally, the begin message puts the broker into relaying mode (line 3), i.e., from now on, new (un)subscriptions received by this broker coming over r are colored gray and relayed to the respective broker connected to a on the same side of the reconfiguration cycle (cf. to the optimization described in Figure 4.7(a)).

The subscription message and the separator message sent (lines 8 and 12) have the relayed flag set, which means that the receiving broker handles them like it had received them over a.

In Figure 4.11, an example is shown, where B_1 sends a begin message m^b to B_1^r and B_2^r. In reaction to receiving m^b, B_1^r and B_2^r send the subscriptions regarding the other side of r to B_1 and B_n in subscription messages m^r, followed by a separator message m^s.

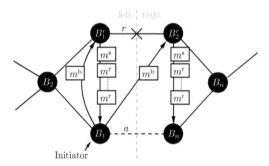

Figure 4.11: Example scenario depicting the case, where B_1 sends begin messages to B_1^r and B_2^r which react by forwarding subscriptions m^r regarding r to $B_{1/2}^a$ followed by a separator message m^s

Separator Message. The name of the separator message is derived from the fact that every broker that receives a separator message changes its color from black to gray. This way, the separator message "separates" the black from the gray messages, because it is guaranteed that no black messages issued on the same side

---

Algorithm 3 Handle begin message m^b

---

1 Set **procedure** handleBeginMsg(Broker B_{sender}, BeginMsg m^b)
2 **begin**
3   relaying ← `true`
4   **for** $(f, B) \in$ RoutingTable **do** // *select subscriptions for the other side of r*
5   **if** $B = B_{old}$ **then**
6   // *send subscription pointing to the other side of r to the respective broker*
7   m^r ← new SubscriptionMsg(filter ← f, color ← gray, relayed ← `true`)
8   sendMsg(B_{relay}, m^r)
9   **endif**
10   **endfor**
11   // *send separator message marking the end of this phase*
12   m^s ← new SeparatorMsg(broker ← ∅, gray ← `false`, relayed ← `true`)
13   sendMsg(B_{relay}, m^s)
14   **if** $B_{old} \in \mathcal{B}_{gray}$ **then** // *reconfiguration already finished on the other side?*
15   sendMsg(B_{relay}, new EndMsg())
16   B_{relay} ←⊥
17   relaying ← `false`
18   **endif**
19   **return** ∅
20 **end**

---

of the reconfiguration cycle, the **separator** message originates from, will follow the **separator** message in the same direction. This is due to the coloring mechanism which colors every message with the color of the first broker on the reconfiguration cycle it encounters.

The brokers B_1^r and B_2^r each issue a **separator** message on their side of the reconfiguration cycle after they have sent all the subscriptions regarding the other side of r to the respective broker at a. The receiving brokers B_1 (B_1^a) and B_n (B_2^a) change their color to gray and forward the **separator** message along their side of the reconfiguration cycle towards the originator of the respective **separator** message.

Every broker that receives a **separator** message which originates from a broker on the same side of the reconfiguration cycle changes its color from black to gray because its routing table is then prepared for the activation of a (the **separator** message is following the subscriptions regarding a). The **separator** message, thus, separates black messages from gray messages—a broker which turned gray colors every notification gray that enters the reconfiguration cycle through it.

A detailed description of the handleSeparatorMsg() procedure is given in Algorithm 4. Every separator message m^s is forwarded along the reconfiguration cycle, i.e., a broker B receiving m^s from B_{left} forwards it to B_{right} and vice versa (line 4). If the message originates from the same side of the reconfiguration cycle the receiving broker changes its color to gray and adds itself to the set of brokers m^s_{brokers} which already turned gray. If B is the first broker at r which receives m^s it also sends a copy of m^s back to the sender in addition to forwarding it over r (line 10) informing the other brokers on its side of the reconfiguration cycle that all brokers have turned gray on this side of the reconfiguration cycle.

The separator message m^s carries the brokers that already turned gray on the side of the initiator of m^s in m^s_{brokers}. On receiving a separator message, B updates its local knowledge about the brokers that have already turned gray (line 14).

The two special cases, where B is connected to r *and* a have to be handled separately. If B is connected to r and m^s has not been sent via r this means that the separator message has passed all brokers on the reconfiguration cycle on the side of B (line 24). Thus, the routing tables on this side of the reconfiguration cycle have been updated for a which means that B can remove all entries stored in its routing table that point to the broker on the other side of r, disabling forwarding over r thereby (line 25).

If B is connected to r, and m^s has been received over r (line 21), accordingly the routing tables on the other side of the reconfiguration cycle are prepared for forwarding messages over a. Thus, B will not receive any black subscriptions over r anymore which would have to be relayed to B_{relay}. Thus, it sends an end message m^e to B_{relay} (i.e., the broker on its side of the reconfiguration cycle connected to a) informing it that it can now start dequeuing the (un)subscriptions in Q_{sub}. Afterwards, the separator message is forwarded on B's side of the reconfiguration cycle to "flush" the channels on this side and finally inform the receiving broker at a that all black notification stemming from the other side of the reconfiguration cycle have left this side.

If B is an endpoint of a and it knows that all necessary brokers (stored in $\mathcal{B}_{\mathrm{queue}}$, depending on the ordering) have received a separator message, it can process the notifications stored in Q_{not} accordingly (line 16). Together with the check on $\mathcal{B}_{\mathrm{queue}}$, this ensures the implementation of Rules 6 and 7 regarding the ordering of notifications ($\mathcal{B}_{\mathrm{queue}}$ has been initialized accordingly in lines 14–19 of Algorithm 2). The brokers connected to a never forward a separator message over a to prevent that the broker on the other side of a receives it for a second time (since it was the first broker which received this separator message). Thus, it drops m^{s} if the next receiving broker would be B_{new} (line 5).

The reconfiguration process is finished from the perspective of B if B is an endpoint of r, its color is gray, and it has received a separator message from the broker on the other side of r (line 32). In this case, it forwards the separator message to the remaining brokers and starts unlocking the reconfiguration cycle by sending an unlock message (line 37). Then, it is finally possible to deactivate the old link by removing the broker on the other side of r from $\mathcal{B}_{\mathrm{neighbors}}$ (line 38). The coloring mechanism together with the FIFO channels between the brokers ensures that no more black messages will arrive later at B which have to be forwarded over r, satisfying Rule 2 and Rule 3 thereby.

Figure 4.12 depicts an example scenario, where B_1 forwards the subscriptions received from B_1^r followed by a separator message m^{s}. In the beginning, m^{s} only carries B_1 in $m^{\mathrm{s}}_{\mathrm{brokers}}$. When it is forwarded by B_2 to B_1^r, it carries the $m^{\mathrm{s}}_{\mathrm{brokers}} = \{B_1, B_2\}$. In the following, when m^{s} is forwarded by B_1^r back to B_1 and over r, the value of $m^{\mathrm{s}}_{\mathrm{brokers}} = \{B_1, B_2, B_1^a\}$ does not change anymore.

End Message. Brokers on r send out an end message m^{e} to the respective broker on the same side of the reconfiguration cycle that is connected to a when they have received a separator message coming over r from the other side of the reconfiguration cycle (line 15 in Algorithm 3). Doing this, they mark the end of the reconfiguration on the other side of the reconfiguration cycle because they will receive no more black messages from there until the reconfiguration has terminated.

Algorithm 4 Handle separator message m^s

---

1 Set **procedure** handleSeparatorMsg(Broker B_{sender}, SeparatorMsg m^s)

2 **begin**

3   *// forwarding of separator message*

4     $\mathcal{M} \leftarrow B_{sender} = B_{right} ? \{(B_{left}, m^s)\} : \{(B_{right}, m^s)\}$ *// forward m^s in direction*

5   **if** $(B_{new}, m^s) \in \mathcal{M}$ **then** $\mathcal{M} \leftarrow \emptyset$ *// do not forward m^s over a*

6   **if** $B_{sender} = B_{relay} \vee (B_{sender} = B_{left} \wedge side = \texttt{left})$

7       $\vee (B_{sender} = B_{right} \wedge side = \texttt{right})$ **then**

8   color $\leftarrow$ **gray** ; $m^s_{brokers} \leftarrow m^s_{brokers} \cup \{B_{this}\}$

9   **if** $B_{old} \neq \perp$ **then**

10         $\mathcal{M} \leftarrow \mathcal{M} \cup \{(B_{sender}, m^s)\}$

11   **endif**

12   **endif**

13   *// update status of brokers that turned gray*

14     $\mathcal{B}_{gray} \leftarrow \mathcal{B}_{gray} \cup m^s_{brokers}$; $\mathcal{B}_{queue} \leftarrow \mathcal{B}_{queue} \setminus m^s_{brokers}$

15   **if** $B_{new} \neq \perp \wedge \mathcal{B}_{queue} = \emptyset$ **then** *// notifications can be dequeued at B_1 or B_n*

16   **for** $(B', m') \in Q_{not}$ **do** *// dequeue in FIFO order*

17         $Q_{not} \leftarrow Q_{not} \setminus (B', m')$

18   handleMsg(B', m')

19   **endfor**

20   **elseif** $B_{old} \neq \perp$ **then** *// broker is connected to the old link*

21   **if** $B_{sender} = B_{old}$ **then** *// m^s has been received over r*

22   sendMsg$(B_{relay}$, new EndMsg())

23         $B_{relay} \leftarrow \perp$; relaying $\leftarrow$ false

24   **else** *// m^s has not been received over r*

25   **for** $(f, B') \in$ RoutingTable **do** *// remove routing entries regarding r*

26   **if** $B' = B_{old}$ **then**

27             $m^n \leftarrow$ new UnsubscriptionMsg(filter $\leftarrow f$, gray $\leftarrow$ false)

28   handleMsg(B_{old}, m^n)

29   **endif**

30   **endfor**

31   **endif**

32   **if** $B_{this} \in \mathcal{B}_{gray} \wedge B_{old} \in \mathcal{B}_{gray}$ **then** *// start unlock process*

33   **for** $(B', m') \in \mathcal{M}$ **do**

34   sendMsg(B', m')

35   **endfor**

36         $m^k \leftarrow$ new UnlockMsg(gray $\leftarrow$ false)

37   handleMsg(B_{old}, m^k)

38         $\mathcal{B}_{neighbors} \leftarrow \mathcal{B}_{neighbors} \setminus \{B_{old}\}$

39         $B_{old} \leftarrow \perp$

40         $\mathcal{M} \leftarrow \emptyset$

41   **endif**

42   **endif**

43   **return** $\mathcal{M}$

44 **end**

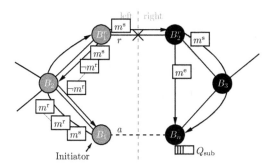

Figure 4.12: Example scenario depicting the case, where B_1^a forwards the set of subscriptions received from B_1^r (according to the routing algorithm used) followed by a **separator** message m^s (B_1^r only forwards m^s over r)

Algorithm 5 details how the receiving broker B processes the **end** message m^e.

Receiving m^e, a broker knows that the sender of m^e will not relay any (un)subscription to him anymore. Thus, B can process the gray (un)subscriptions stored in Q_{sub} (line 4) that have been received during reconfiguration and have been received over a. This procedure implements Rule 5.

When B receives an **end** message it may already have received an unlock message that has been processed but not forwarded because of the missing **end** message (line 8). This is possible due to varying message delays. In this case, B has to take care of activating a by sending an **unlock** message to itself (line 11).

Unlock Message. Complementary to the **lock** message which locks the brokers on the reconfiguration cycle to prevent race conditions due to parallel overlapping reconfigurations, the unlock message m^k is responsible of freeing the brokers after the reconfiguration is implemented. A broker B which receives m^k initializes its pointers according to the direction from which m^k has been received (line 12 in Algorithm 6). The broker is finally unlocked if it has received an **unlock** message from both neighbors on the reconfiguration cycle (and an **end** message if the broker is connected to a, line 16).

Algorithm 5 Handle end message m^{e}

---

1 Set **procedure** handleEndMsg(Broker B_{sender}, EndMsg m^{e})
2 **begin**
3    $B_{\mathsf{relay}} \leftarrow \perp$ // *reconfiguration finished on the other side*
4   **for** $(B', m') \in Q_{\mathsf{sub}}$ **do** // *process queued (un)subscriptions*
5      $Q_{\mathsf{sub}} \leftarrow Q_{\mathsf{sub}} \setminus (B', m')$
6   handleMsg(B', m')
7   **endfor**
8   **if** $(B_{\mathsf{left}} = B_{\mathsf{new}} \wedge B_{\mathsf{right}} = \perp) \vee (B_{\mathsf{right}} = B_{\mathsf{new}} \wedge B_{\mathsf{left}} = \perp)$
9      $\vee \, (B_{\mathsf{left}} = \perp \wedge B_{\mathsf{right}} = \perp)$ **then** // *received unlock message meanwhile*
10     $m^{\mathrm{k}} \leftarrow$ new UnlockMsg(gray $\leftarrow$ false)
11   handleMsg$(B_{\mathsf{this}}, m^{\mathrm{k}})$
12   **endif**
13   **return** $\emptyset$
14 **end**

Sending an initial unlock message is triggered on the receipt of a separator message at brokers connected to r (cf. Algorithm 4, line 37). In this case, B sends an unlock message to itself which is then handled like it had been received from the other side of r and is forwarded in both directions on the reconfiguration cycle.

A broker B that receives m^{k} for the first time forwards it to both neighbors on the reconfiguration cycle (line 3), where the message to the original sender serves as an acknowledgement which finally unlocks the broker (line 7). An unlock message that has been received for the first time by a broker connected to a will neither be acknowledged nor forwarded if B has not yet received an end message (line 11). If an unlock message *and* an end message has been received by a broker connected to a, the unlock message will be forwarded over a to finally activate it (line 5).

It is not necessary to send m^{k} over r because the unlock process is already triggered by a separator message for $B^r_{1/2}$ (line 10).

Figure 4.13 depicts an example scenario, where B^r_1 sends an end message m^{e} to B_1 and an unlock message m^{k} to B_2. B_2 acknowledges m^{k} and forwards it to B_1 which also acknowledges it and additionally forwards it to B_n. B_n does not forward m^{k} until it received an end message m^{e} from B^r_2 and a separator message

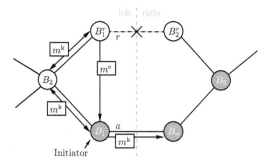

Figure 4.13: B_1^r sends an unlock message m^k to B_1 and B_2. Both forward them counter clockwise on the reconfiguration cycle (B_2 and B_1 acknowledge m^k with another unlock message)

from B_3.

Modifications of Basic Operations

The advanced coordinated reconfiguration algorithm relies on queueing and coloring of messages. Up to now, only the different new message types introduced to coordinate the reconfigurations have been addressed and coloring and queueing issues have been left aside. The latter are addressed in the procedures receiveMsg() (Algorithm 7) and sendMsg() (Algorithm 8) described in Figure 4.9 and presented in the following.

Receiving a Message. When broker B receives a message, it may need to be delayed due to ordering requirements. This affects notifications and (un)subscriptions received over a. Notifications have to be delayed in case of ordering requirements and are, thus, enqueued accordingly (Algorithm 7, line 3). (Un)subscriptions have to be delayed if B has not yet received an end message from B_{relay} due to FIFO-publisher ordering requirements (line 5).

If B is connected to r, in relaying mode, and an (un)subscription has been received over r, this (un)subscription is relayed to B_{relay} (line 7). This is done to

---

Algorithm 6 Handle unlock message m^k

---

1 Set **procedure** handleUnlockMsg(Broker B_{sender}, UnlockMsg m^k)

2 **begin**

3   **if** $B_{\text{sender}} \neq B_{\text{new}} \wedge B_{\text{left}} \neq \perp \wedge B_{\text{right}} \neq \perp$ **then** // *received first unlock message*

4     $\mathcal{M} \leftarrow \{(B_{\text{left}}, m^k), (B_{\text{right}}, m^k)\}$

5   **elseif** $B_{\text{new}} \neq \perp \wedge (B_{\text{left}} = \perp \vee B_{\text{right}} = \perp)$ **then** // *2nd unlock message at a*

6     $\mathcal{M} \leftarrow \{(B_{\text{new}}, m^k)\}$

7   **else** // *received an acknowledgement for an unlock message*

8     $\mathcal{M} \leftarrow \emptyset$

9   **endif**

10   $\mathcal{M} \leftarrow B_{\text{sender}} = B_{\text{old}} ? \mathcal{M} \setminus \{(B_{\text{old}}, m^k)\} : \mathcal{M}$ // *do not send over r*

11   $\mathcal{M} \leftarrow B_{\text{relay}} \neq \perp ? \mathcal{M} \setminus \{(B_{\text{new}}, m^k)\} : \mathcal{M}$ // *wait for end message*

12   $B_{\text{left}} \leftarrow B_{\text{sender}} = B_{\text{left}} ? \perp : B_{\text{left}}$ // *forward on the reconf. cycle*

13   $B_{\text{right}} \leftarrow B_{\text{sender}} = B_{\text{right}} ? \perp : B_{\text{right}}$ // *forward on the reconf. cycle*

14   $B_{\text{new}} \leftarrow (B_{\text{new}}, m^k) \in \mathcal{M} ? \perp : B_{\text{new}}$

15   color $\leftarrow B_{\text{new}} = \perp ?$ black $:$ color // *change color back to default*

16   **if** $B_{\text{left}} = \perp \wedge B_{\text{right}} = \perp \wedge B_{\text{new}} = \perp$ **then** // *unlock process finished*

17    $\mathcal{B}_{\text{gray}} \leftarrow \emptyset$

18   locked $\leftarrow$ false

19   **endif**

20   **return** $\mathcal{M}$

21 **end**

Algorithm 7 Receiving a message

1 void **procedure** receiveMsg(Broker B_{sender}, Msg m)
2 **begin**
3   **if** $B_{\text{sender}} = B_{\text{new}} \wedge \mathcal{B}_{\text{queue}} \neq \emptyset \wedge m$ is notification **then**
4   enqueue(Q_{not}, (B_{sender}, m)) // *delay notifications coming over a*
5   **elseif** $B_{\text{sender}} = B_{\text{new}} \wedge B_{\text{relay}} \neq \perp \wedge m$ is (un)subscription **then**
6   enqueue(Q_{sub}, (B_{sender}, m)) // *delay unil end message received*
7   **elseif** $B_{\text{sender}} = B_{\text{old}} \wedge$ relaying $\wedge m$ is (un)subscription **then**
8     $m_{\text{relayed}} \leftarrow$ true
9   sendMsg(B_{relay}, m)
10   **else**
11   **if** m_{relayed} **then** // *message has been relayed*
12       $m_{\text{relayed}} \leftarrow$ false ; $B_{\text{sender}} \leftarrow B_{\text{new}}$
13   **endif**
14   **if** $B_{\text{sender}} \neq B_{\text{left}} \wedge B_{\text{sender}} \neq B_{\text{right}} \wedge$ color = gray **then**
15       $m_{\text{gray}} \leftarrow$ true // *color message on reconfiguration cycle entry*
16   **endif**
17   handleMsg(B_{sender}, m)
18   **endif**
19 **end**

implement the optimization proposed for Rule 5 as depicted in Figure 4.7(a). The receiving broker of a relayed message handles it like a regular message that has been received over a (line 11).

If B is the first broker on the reconfiguration cycle that received a message m and B is gray, then m is colored accordingly (line 14). Messages which have their gray-flag set are interpreted as being gray. Otherwise, they are handled as black messages.

Messages that are not queued or relayed are handled in the conventional way by using them as a parameter when calling handleMsg() procedure (line 17).

Sending a Message. Similarly to the extra pre-processing of messages when receiving them, it may be necessary to post-process a message in the sendMsg() procedure before it is sent (Figure 4.9). This affects the case, where messages have to be "uncolored" when they leave the reconfiguration cycle (i.e., they are colored with the default color black, line 6). If the message is sent to the broker itself, the handleMsg() procedure is directly called (line 8).

Gray messages which should be sent via r and black messages which should be sent over a are discarded silently except for lock, unlock, begin, and end messages (line 3). The other message types discussed above are handled regularly because they are necessary for the reconfiguration algorithm and are, thus, not forwarded according to their color. Dropping notifications and (un)subscriptions with the wrong color is necessary to avoid duplicates as required in Rule 1 and Rule 4.

Algorithm 8 Sending a message

```
 1  void procedure sendMsg(Broker Breceiver, Msg m)
 2  begin
 3    if  (m is lock, unlock, begin, or end message)
 4         ∨ ¬ ((Breceiver = Bnew ∧ ¬mgray) ∨ (Breceiver = Bold ∧ mgray)) then
 5       if  Breceiver ∉ {Bthis, Bleft, Bright, Bnew, Bold, Brelay} then
 6          mgray ← false // remove color if receiver not on the reconfiguration cycle
 7       endif
 8       if  Breceiver = Bthis then // message is sent to this broker
 9          receiveMsg(Breceiver, m)
10       else
11          send(Breceiver, m)
12       endif
13    endif
14  end
```

Advertisements

Advertisements are used to build routing tables for subscriptions. Accordingly, the same rules apply to them as for (un)subscriptions, i.e., it is necessary to ensure a FIFO-producer ordering for them. Advertisements must be processed before the subscriptions such that the subscription routing tables are already built when the subscriptions are forwarded. Thus, advertisements can be incorporated just like subscriptions. However, it has to be taken care that the advertisements are processed *before* the subscriptions.

4.4.3 Evaluation

This section presents the results of experiments conducted using a discrete event simulation. The goal of the experiments was to measure the performance of the presented algorithm. The performance can be measured regarding the overhead of the algorithm and regarding the message delay induced during reconfiguration. The latter is necessary to provide message ordering guarantees as described earlier and induces extra delays compared to an algorithm which does not give ordering guarantees and, thus, does not need to delay any message.

The evaluation focuses on the average delay of a notification in the queue Q_{not} managed by the brokers connected to a. Furthermore, additional messages have to be sent for coordination purposes which leads to a higher bandwidth consumption. All measurements are compared with the results gained when applying the basic STRAWMAN approach that has been discussed as the most basic uncoordinated approach in Section 4.4.1. Moreover, the number of duplicates and notifications lost are measured when applying the STRAWMAN approach which gives an impression of the impact of missing guarantees. In the experiments, the focus is on the interesting case of exchanging a link in the broker topology.

Simulation Setup

The simulation scenario is chosen in a way such that sufficiently complex message streams are created to evaluate the new algorithm with respect to correctness and overhead. The number of brokers is chosen such that reconfiguration cycles of sufficient sizes can be evaluated. The simulated network relies on an Internet-like physical topology to obtain realistic delay times.

All experiments are based on a transit-stub network topology with a total number of 10,000 nodes, subdivided into 100 domains of equal size. The topology was generated using the BRITE topology generator [109] with the configuration described in Appendix A.1. The publish/subscribe broker overlay topology is composed of 100 randomly chosen brokers which initially form a randomly generated acyclic graph. In the broker overlay network, every broker can potentially connect to any other broker. The maximum delay of an overlay link is limited to 100 ms, while a broker needs at most 0.01 ms to process a message (in this scenario, one tick of the discrete event simulation corresponds to 10 ms). The brokers host a total of 500 clients, which are uniformly distributed among them. The clients consist of 50 publishers, each having a set of 9 dedicated subscribers that receive all of the notifications it publishes.

The publication rate follows an exponential distribution and a publisher creates a new notification every 50 ms on average. The following experiments were repeated up to 50 times and arithmetic means as well as 95% confidence intervals were calculated for every measured value. Thus, 50 simulation runs were executed per point measured in each experiment. To ensure that results of different algorithms are comparable, the same setups and random seeds were used for corresponding simulation runs.

Reconfiguration Overhead

In the first experiment, control message complexity of the advanced coordinated reconfiguration algorithm is evaluated and compared to the STRAWMAN approach. Therefore, 100 randomly chosen links are sequentially exchanged such that the network stays connected and acyclic, while the control messages which are necessary to update the brokers' routing tables are counted.

Figure 4.14 shows the number of control messages as a function of the length of the reconfiguration cycle. The message complexity increases for both algorithms because (un)subscriptions have to be forwarded to more brokers to update their routing tables if the reconfiguration cycle grows in size. Since the advanced coordinated reconfiguration algorithm limits filter forwarding to the reconfiguration cycle, it clearly outperforms the STRAWMAN approach especially for smaller sizes. Although the number of control messages increases due to additional messages needed for coordination, the advanced coordinated reconfiguration algorithm remains more efficient even for bigger reconfiguration cycles. Please note that the

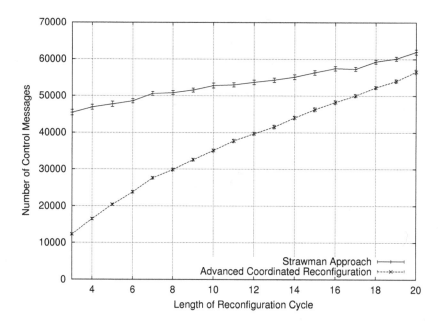

Figure 4.14: Number of control messages versus length of reconfiguration cycle

overhead induced by the STRAWMAN approach is only limited by the size of the topology and can, thus, grow with a growing number of brokers—independent of the length of the reconfiguration cycle. In contrast to this, the overhead of the advanced coordinated reconfiguration algorithm is limited by the size of the re- configuration cycle. The overhead of both algorithms, of course, also depends on the number of (un)subscriptions issued by the clients.

The results of this experiment show that the advanced coordinated reconfigura- tion algorithm is especially well suited to shorter reconfiguration cycles since the message overhead is reasonably low then. The additional overhead of the STRAW- MAN approach also increases with the length of the reconfiguration cycle although no extra messages for coordination are induced. This is due to the fact that with a growing reconfiguration cycle length, it takes longer to send (un)subscriptions from the endpoints of a to the endpoints of r which increases the chance of (un)subscriptions leaving the reconfiguration cycle and causing additional overhead thereby.

Message Loss and Duplicates

The prevention of message loss and duplicates during reconfiguration is one of the most important features that distinguishes the advanced coordinated reconfiguration algorithm from the other reconfiguration algorithms in literature. The second important feature this algorithm offers is that of providing ordering guarantees during reconfiguration. Before continuing with an experiment which measures the message delay that is induced by the reconfiguration algorithm in order to give ordering guarantees, some results regarding lost notifications and duplicates received when using the STRAWMAN approach are presented. The aim is to illustrate the quantity of notifications that are lost or received more than once by a broker depending on the reconfiguration cycle length.

The experiment builds on the same settings as the previous experiment. The number of notifications lost and duplicates received increases with a growing reconfiguration cycle as shown in Figure 4.15, where the number of duplicates received and the number of notifications lost is approximately linear in the number of brokers on the reconfiguration cycle. The number of lost notifications is far from being negligible, showing that this is a severe issue in the STRAWMAN approach.

Message Delay

The advanced coordinated reconfiguration algorithm prevents message loss during reconfiguration, but introduces a delay as notifications are queued after passing the new link a, in case a certain ordering has to be guaranteed. The delay depends on the type of ordering (FIFO-publisher or causal) and on the length of the reconfiguration cycle. The goal of this experiment is to explore the impact of the type of ordering guarantees and the size of the reconfiguration cycle on the

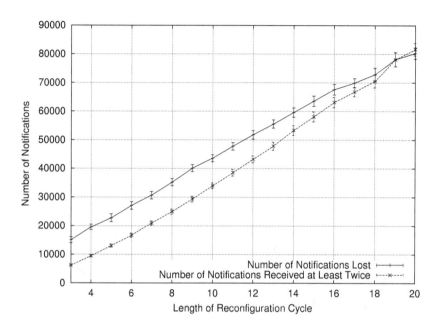

Figure 4.15: Number of notifications lost and number of duplicates received with the STRAWMAN approach

average delay of a notification.

Figure 4.16 shows the average delay induced by 100 random link exchanges measured as the mean time an affected notification is queued at the brokers on the new link. The delay increases linearly with the length of the reconfiguration cycle. This is due to the fact that **separator** messages have to travel over the reconfiguration cycle before queued notifications are released again. Causal ordering introduces more delay than FIFO-publisher ordering because notifications are queued until **separator** messages have arrived from both sides of the reconfiguration cycle in this case. With FIFO-publisher ordering, however, notifications are already dequeued on the receipt of one **separator** message from the opposite side of the reconfiguration cycle.

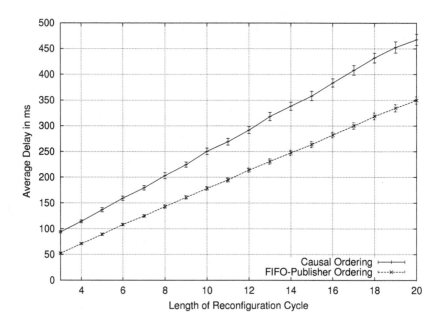

Figure 4.16: Average delay caused by ordering versus length of reconfiguration cycle

4.4.4 Discussion

The results of the experiments show that the issue of notification loss and reduction of message overhead are handled well by the advanced coordinated reconfiguration algorithm. If, however, ordering guarantees are needed, an additional delay is introduced which grows with the size of the reconfiguration cycle. By comparing it with the STRAWMAN solution, a common approach in literature [50, 64, 126] is taken although a comparison is difficult because the STRAWMAN approach produces duplicates and notification loss. Similarly, all other discussed reconfiguration algorithms for publish/subscribe systems may produce notification loss and do not provide any ordering guarantees at all. In [64], Frey presents a comparison of the algorithms discussed in Section 4.4.1 with the STRAWMAN approach regarding message loss and overhead.

The results also show that the advanced coordinated reconfiguration algorithm produces the least overhead for short reconfiguration cycles. This is due to the forwarding of (un)subscriptions from the brokers connected to r and a, respectively.

The algorithm presented needs to lock the reconfiguration cycle before the actual reconfiguration can start because parallel overlapping reconfigurations are not supported. If multiple parallel overlapping reconfigurations all start locking their reconfiguration cycle with an unfavorable timing and, thus, encounter themselves at different brokers, it is possible that a livelock is produced if all brokers restart the reconfiguration over and over again. This problem can be avoided probabilistically if restarts are delayed for a random time period after a failed try to lock the reconfiguration cycle.

4.5 Reconfiguring Self-Stabilizing Publish/Subscribe Systems

In the previous section, an algorithm for reconfiguring broker overlay networks in conventional publish/subscribe systems is presented, where a fault-free environment is assumed and brokers have direct access to the topology management. Besides efficiency, the focus was on the prevention of message loss and duplicates, and the provision of ordering guarantees. This section ties up to Chapter 3, where self-stabilizing content-based routing is introduced for publish/subscribe systems. Self-stabilizing routing is complemented with a self-stabilizing broker overlay network to create a self-stabilizing publish/subscribe system with respect to content-based routing and the broker overlay layer. According to the reconfiguration algorithm for conventional publish/subscribe systems, the focus here is on elementary reconfigurations as introduced in Section 4.2.2. An algorithm is proposed which is capable of reconfiguring the broker topology without message loss. The layered algorithm stack presented is able to cope with several problems regarding the reconfiguration of a self-stabilizing publish/subscribe system.

This section starts with a short survey of how reconfigurations of self-stabilizing systems are handled in related work in Section 4.5.1 and points out some of the basic problems in this context. Then, the general challenges which have to be solved when reconfiguring layered self-stabilizing publish/subscribe systems are discussed in Section 4.5.2. Section 4.5.3 presents the new solution in detail. Section 4.5.4 presents extensions to incorporate causal and FIFO-publisher ordering. Further extensions and other issues are discussed in Section 4.5.5.

4.5.1 Related Work

The power of self-stabilizing systems lies in their ability to bring themselves into a legitimate state when started in an arbitrary configuration without any manual intervention if no fault occurs for a long-enough time period. This ability results from the convergence property of self-stabilizing algorithms. The closure property further provides the guarantee that a system in a legitimate state stays there as long as no fault occurs. However, during stabilization, no guarantees on system behavior are given.

When a system is reconfigured, i.e., it is transformed from one legitimate configuration to another one, it is intuitively expected that certain service guarantees hold. This is in contrast to the situation in which a sudden fault makes it necessary to reconfigure the system. In the following it is shown that this expectation is not always fulfilled for self-stabilizing systems due to subtle difficulties. As a starting point, an overview of different approaches to reconfiguration of self-stabilizing systems in literature is given in the following.

Handling Reconfigurations as Faults

The naive approach which is implicitly taken by many layered self-stabilizing systems that do not consider reconfigurations explicitly, is to handle reconfigurations as faults from which the system recovers eventually. For a publish/subscribe system this could mean that changes in the topology of the broker overlay network result in inconsistencies on the routing layer because it is closely related to the topology of the broker overlay. Due to the self-stabilizing mechanism used, the routing tables of the brokers stabilize eventually as described in Chapter 3.

This approach does not need any coordination efforts on the cost of possible message loss or duplicates because the broker overlay network is reconfigured regardless of the routing tables of the brokers. Figure 4.17(a) depicts how a layered self-stabilizing system is expected to react in case of a reconfiguration on one layer, while Figure 4.17(b) depicts what may actually happen (as demonstrated in the example) if special care has not been taken.

This problem does not only occur in layered self-stabilizing systems although the problems are obvious here. In many self-stabilizing systems that do not handle reconfigurations explicitly, a reconfiguration may also lead to an illegitimate state from which the system will recover eventually if the reconfiguration rate does not pass a given frequency threshold. This can happen, for example, if a single reconfiguration like a link exchange in a topology may demand for other reconfigurations that are caused as a side-effect by the self-stabilizing algorithm. In these cases, special measures have to be taken to not interrupt the service of a self-stabilizing system due to reconfigurations. This issue will be discussed in

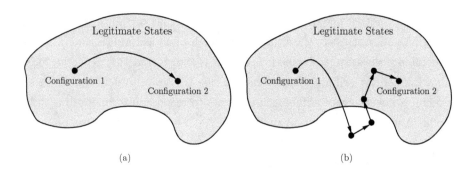

Figure 4.17: Effect of reconfigurations in a self-stabilizing system as (a) expected
and (b) possible without special treatment

greater depth in the following and in Section 4.5.3.

Fault Containment

As already mentioned, a serious problem with the naive approach is that reconfiguring a self-stabilizing system may lead to diverse side-effects because reconfiguring
a lower layer may affect all upper layers. Take a spanning tree, for example, which
is self-stabilizing and aligns its structure according to the node IDs. In this case, a
reconfiguration of the topology by adding a new node with the highest (or lowest)
ID can lead to a change of the whole tree topology in the worst case which again
may demand for a reconfiguration of the routing tables of the brokers.

Ghosh et al. propose a way to limit the side effects of certain faults in a self-
stabilizing system by introducing the concept of *fault containment* [68]. However,
the goals of fault containment and self-stabilization are conflicting in general since
adding fault containment to a self-stabilizing protocol may increase the stabilization time [69]. Although fault containment limits the side-effects of reconfigurations which may lead a fault on an upper layer, it does not prevent service
interruption which may constitute in lost messages. This is due to the fact that
still no special care is taken of reconfigurations. However, fault containment may

reduce the impact of a reconfiguration on the system during convergence.

Superstabilization

Dolev and Herman [58] propose the class of *superstabilizing* protocols which are self-stabilizing and explicitly designed to cope with topology dynamics. Supersta- bilization builds on the idea that a topology undergoes "typical" changes which are subsumed in the class Λ of change events. A *passage predicate* is introduced which is weaker than the predicate specifying the legitimate states but still use- ful. In case a topology change event defined in Λ happens, the passage predicate provides certain guarantees until the system is back in a legitimate state.

As an example, the authors propose a superstabilizing spanning tree algorithm, where the removal of network links that are not part of the spanning tree and link exchanges do not lead to a rebuild of the topology. This algorithm is still self-stabilizing and cycles in the topology are detected and corrected eventually.

Superstabilization represents an effort which explicitly addresses the issue of reconfiguration. Although very interesting, it does not provide a general solution to create superstabilizing protocols. Moreover, the case of layered self-stabilizing algorithms is not discussed, thus, limiting its usefulness for the problem domain targeted here.

4.5.2 Challenges

The conventional approaches to reconfigure a (layered) self-stabilizing system pre- sented above already indicated the challenges that are inherent when reconfiguring the broker overlay topology of a self-stabilizing publish/subscribe system. In the following, they are described in detail with respect to self-stabilization and, thus, supplement the challenges already described in Section 4.3. Thereby, it is assumed that a reconfiguration transforms a broker overlay topology from configuration $\mathfrak{T}_1$ to another configuration $\mathfrak{T}_2$.

No Service Interruption

During and after reconfiguration, the system must stay in a legitimate state given that no fault occurs. For a self-stabilizing publish/subscribe system this means that the correctness of the publish/subscribe system according to Definition 5 must be maintained which implies that messages must not be lost and clients must not receive notifications more than once.

Containment of Changes

A reconfiguration must not result in unintended additional reconfigurations that are not desired by the administrator executing the reconfiguration by purpose. For the broker overlay network this means that one elementary reconfiguration (e.g., a link exchange) must not result in an unwanted link exchange that was not explicitly requested.

At this point, it might appear sensible at first sight to simply apply the advanced coordinated reconfiguration algorithm from Section 4.4.2 for the superstabilizing spanning tree algorithm described above together with the self-stabilizing content-based routing algorithms presented in Section 3. However, this results in problems because the reconfiguration algorithm creates a cyclic broker overlay topology for a limited time period during reconfiguration (cf. Algorithm 2 on page 137, line 29) which would be handled as a fault by the superstabilizing algorithm. Besides that, the superstabilizing algorithm requires a "neighborhood" between the brokers in terms of brokers that can act as neighbor brokers which is not available by default.

Persistence of Changes

Many self-stabilizing algorithms imply a certain structure that is legitimate and towards which the algorithm "pushes" the system. If the reconfiguration does not comply with this structure, it is "repaired" and possibly undone thereby.

An illustrative example is the case, where a self-stabilizing spanning tree algorithm is used for maintaining the broker overlay network. Assume that the self-stabilizing tree algorithm by Afek, Kutten, and Yung is used [3]. Its aim is to build a spanning tree, where every node has a unique identifier (*ID*) and eventually the node with the biggest ID becomes the root. Every other node connects either directly to the root node or, if this is not possible, to the node which is closest to the root (in terms of hops) with the biggest ID (if more than one node has the minimum distance to the root).

An example graph with all possible overlay links is depicted in Figure 4.18(a). Here, every node runs the self-stabilizing tree algorithm. After a while, the topology stabilizes to the configuration depicted in Figure 4.18(b). If one now reconfigures the topology in a way such that the link between the nodes 8 and 3 is exchanged with the link between nodes 3 and 2 (Figure 4.18(c)), this will be considered as a fault by the self-stabilizing tree algorithm since node 3 is not connected to the root node by the shortest path anymore. Thus, the topology will stabilize again to the one depicted in Figure 4.18(b) and the reconfiguration is lost.

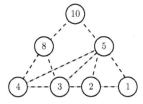 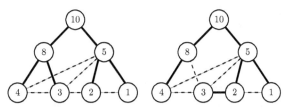

(a) Initial configuration, where the numbers resemble the node IDs and the dashed lines depict communication links between nodes

(b) Configuration after the algorithm stabilized (the solid lines depict the links of the spanning tree)

(c) Configuration after exchanging the link $\overline{83}$ with $\overline{32}$.

Figure 4.18: Example for reconfiguring a self-stabilizing spanning tree topology according to [3]

In this example, it is obviously not possible to integrate arbitrary reconfigurations without adapting the broker IDs accordingly. It is hence a good example for the subtle problems that can arise when trying to integrate reconfiguration into self-stabilizing systems.

4.5.3 Coordinated Reconfiguration with Layered Self-Stabilization

In Chapter 3, algorithms were presented that render the content-based routing layer self-stabilizing. Thereby, it is assumed that the broker overlay network is either static or self-stabilizing. Although a static broker overlay makes the whole system formally self-stabilizing it is not feasible in practice because reconfigurations of the topology might become necessary due to system dynamics. It has already been argued that reconfiguring a self-stabilizing publish/subscribe system is different from reconfiguring a regular one. Thus, a self-stabilizing broker overlay network is proposed in this section on which self-stabilizing content-based routing is layered upon and a coordination mechanism between the overlay and the routing layer is introduced to cope with the challenges described above.

Systems which are layered like in this case here can be made self-stabilizing by making all layers individually self-stabilizing. This transparent stacking of self-stabilizing layers is a standard technique which is referred to as *fair composition* [57]. It is easy to combine self-stabilizing algorithms this way to create a new and more powerful self-stabilizing mechanism as long as no cyclic dependencies exist among their states. Taking this approach, it is sensible to layer self-stabilizing routing in publish/subscribe systems on top of a self-stabilizing broker topology which employs a self-stabilizing tree algorithm like the ones given in literature (cf. Gärtner's survey [66] for a good overview). However, this approach has its drawbacks as already discussed.

The approach in the following is, hence, to realize a self-stabilizing overlay topology which maintains an arbitrary tree structure and to layer self-stabilizing routing on top of it in a way, such that reconfigurations of the overlay topology can be processed without service interruption (i.e., message loss or duplication). Two problems have to be tackled to solve this problem: (i) designing a self-stabilizing broker overlay network which does not necessarily impose a certain structure on the resulting tree and accepts all correct trees and (ii) coupling the self-stabilizing mechanisms on the overlay and the routing layer to allow for atomic topology switches without message loss.

Coloring Scheme

Before continuing with the description of the self-stabilizing broker overlay network, a coloring scheme is introduced which is used to synchronize reconfigurations on the overlay layer with reconfigurations on the routing layer. To achieve this, selected data structures are marked with a *color* attribute. On the overlay topology layer this concerns the child and parent broker pointers, while on the routing layer the routing entries are affected. Thus, each color may represent a different topology.

To make atomic switches between different colors (and, thus, topologies) possible, every broker maintains data structures for three different colors which can be accessed on the respective layer and are stored in the following variables. The variable c^{cur} stores the color of the topology that is currently used, c^{old} stores the color of the topology that has been used last, and c^{new} stores the color of the topology that will be used when the color changes for the next time. The values of these variables are rotated regularly. Three different colors are required because the communication and processing delays in the network can lead to a situation, where messages are still forwarded although the topology has changed meanwhile. If the value of c^{cur} becomes the value of c^{old}, for example, there may still be messages on the network that are colored with c^{old}. To be able to deliver these messages, the topology for c^{old} has to be kept sufficiently long. In the following, without loss of generality it is assumed that the routing entries are stored in separate routing tables for each color although a tag on each entry suffices in the implementation.

It is the task of the root broker R to regularly recolor all brokers in the tree. To accomplish this, a timeout runs on R. In order to realize self-stabilization, the timeout also runs on every broker $B \neq R$. The actions taken on a timeout are described in Algorithm 9 and will be explained in the following. After resetting the timer (line 2), the root broker rotates its colors and initializes the child broker pointers $\mathcal{C}^{c^{\mathrm{new}}}$, the parent broker pointer $\mathcal{P}^{c^{\mathrm{new}}}$, and the routing table $T^{c^{\mathrm{new}}}$ (lines 5 and 6). Then, the new color is disseminated in a *recolor message* $\mathrm{REC}_{\mathrm{msg}}$ to all child brokers in the topology with the current color. If a child broker does not acknowledge this message, it will be removed (procedure sendToChildren() in Algorithm 12).

Algorithm 9 Procedure onTimeout() which is called regularly on every broker

1 **procedure** onTimeout()
2   resetTimer()
3   **if** $B = R$ **then**
4   // *root broker starts a new coloring period*
5        $c^{\mathrm{old}} \leftarrow c^{\mathrm{cur}}; \; c^{\mathrm{cur}} \leftarrow c^{\mathrm{new}}; \; c^{\mathrm{new}} \leftarrow c^{\mathrm{cur}} + 1 \bmod 3$
6        $\mathcal{C}^{c^{\mathrm{new}}} \leftarrow \mathcal{C}^{c^{\mathrm{cur}}}; \; \mathcal{P}^{c^{\mathrm{new}}} \leftarrow \mathrm{null}; \; T^{c^{\mathrm{new}}} \leftarrow \mathrm{init}$
7        $m \leftarrow$ new $\mathrm{REC}_{\mathrm{msg}}(c^{\mathrm{new}}, \mathfrak{R})$
8   applyReconfig($\mathfrak{R}$)
9        $\mathfrak{R} \leftarrow \emptyset$
10   sendToChildren(m)
11   **else** // *every other broker reconnects to the tree*
12   joinTree ()
13   **endif**

If another broker that is not the root broker runs into a timeout, this is related to a fault because the timers are chosen in a way such that every broker except for the root broker will never experience a timeout if no fault happens. The broker hence tries to reconnect to the tree (line 12). Here, the coloring mechanism is used to realize a self-stabilizing overlay topology. Details on this topic are presented in the following section. The variable $\mathfrak{R}$ carries the reconfigurations to be implemented. They are disseminated together with the recolor message and will be explained in detail later.

When a broker $B \neq R$ receives a recolor message as described in Algorithm 10, it resets its timer, replies with an *acknowledge message* $\mathrm{ACK}_{\mathrm{msg}}$, rotates its colors (and changes the value of its current color c^{cur} thereby), and forwards the message to its child brokers.

The broker accepts the recolor message only if it has been sent by the bro-

---

Algorithm 10 Procedure onReceiveRec() which is called when a recolor message is received

---

1 **procedure** onReceiveRec(REC$_{msg}$ m)
2   **if** sender $= \mathcal{P}^{c^{new}}$ **then**
3   resetTimer()
4   Send ACK$_{msg}$ to sender
5     $c^{old} \leftarrow c^{cur}; c^{cur} \leftarrow nColor; c^{new} \leftarrow m.c$
6     $\mathcal{C}^{c^{new}} \leftarrow \mathcal{C}^{c^{cur}}; \mathcal{P}^{c^{new}} \leftarrow \mathcal{P}^{c^{cur}}; T^{c^{new}} \leftarrow$ init
7   applyReconfig($m.\mathfrak{R}$)
8   sendToChildren(m)
9   **endif**

---

ker $\mathcal{P}^{c^{new}}$ points to (line 2). This test is needed to detect cycles that may result from faults. It might happen that the parent/child pointers of some brokers are perturbed in a way such that a cycle is created in the tree, for example, if the parent pointer of a broker B points to an ancestor broker B', which also has a child pointer to B. In this case it is not obvious from the local view of the brokers that a cycle exists and recolor messages would be forwarded in the cycle forever if B' would accept recolor messages from B. Since B' does not accept and acknowledge a recolor message from B, B will eventually remove B' from his set of child brokers as described in Algorithm 12. Tree partitions are detected since recolor messages are not received in the partitions which do not contain R. Thus, the brokers in the partition will eventually run in a timeout. The rest of the procedure is similar to the procedure onTimeout() in Algorithm 9 except for replying with an acknowledge message to the sender to indicate that the broker is alive (line 4).

The receiver of an acknowledge message sets a flag for the sending broker as described in Algorithm 11.

Algorithm 11 Procedure onReceiveAck()

---

1 **procedure** onReceiveAck(ACK$_{msg}$ m)
2   **if** sender $\in \mathcal{C}^{c^{cur}}$ **then**
3   Set flag (sender)
4   **endif**

This flag is used for a second chance algorithm in the procedure sendToChildren() to remove faulty child broker pointers as previously discussed for the case of cycles (Algorithm 12, lines 3–8).

Algorithm 12 Procedure sendToChildren()

---

1 **procedure** sendToChildren(REC_{msg} m)
2 **foreach** $B' \in C^{c^{cur}}$ **do**
3   **if** flag (B') is set **then**
4   Send m to B'
5   Unset flag (B')
6   **else**
7   Remove child broker B'
8   **endif**
9 **endfor**

Timeouts on the Broker Overlay Layer

The dissemination of recolor messages has already been discussed and parts of the overlay network management have been introduced, where faulty child broker pointers are detected and discarded. In this section, timeouts are further specified and the integration of brokers which were disconnected from the overlay due to a fault is presented.

The self-stabilizing mechanism of the broker overlay network is based on time-outs regarding the receipt of recolor messages. On R, a timeout triggers a new recolor message to be sent to all its child brokers. On every other broker, a timeout occurs if the broker has not received a correct recolor message in time. As recolor messages are forwarded recursively down the tree, the last leaf broker receives the message at the latest after time $h \cdot \delta_{\max}$, where h is the maximum height of the tree (i.e., the diameter d of the graph is at most $2 \cdot h$) and $\delta_{\max}$ ($\delta_{\min}$) is the maximum (minimum) delay for processing and sending a message to a child broker. As the tree may degenerate arbitrarily h can be at most equal to the maximum number of brokers η in the system (which is assumed to be known and stored in ROM for convenience). Given that the timeout on R occurs every time period ξ, a timeout

ξ' with

$$\xi' = \xi + h \cdot (\delta_{\max} - \delta_{\min}) \qquad (4.1)$$

is necessary on every broker B distinct from R, which is resetted every time a new recolor message is received from its parent broker (Algorithm 10, line 3). Algorithm 13 describes the details of the resetTimer() procedure.

Algorithm 13 Procedure resetTimer()

---

10 **procedure** resetTimer()
11   **if** $B = R$ **then**
12   timer $\leftarrow \xi$
13   **else**
14   timer $\leftarrow \xi + h \cdot (\delta_{\max} - \delta_{\min})$
15   **endif**

When $B \neq R$ runs into a timeout, it took more than ξ' to receive the next recolor message after the last one. This can only be due to a fault, since forwarding a message from R to B cannot take more than $h \cdot \delta_{\max}$. In this case, B tries to rejoin the tree which is described in Algorithm 14.

Algorithm 14 Procedure joinTree()

---

16 **procedure** joinTree()
17   nxtParent $\leftarrow R$
18   **do** // ask recursively for a new parent
19   tryParent $\leftarrow$ nxtParent
20   nxtParent $\leftarrow$ tryParent.askForPlaceOrPointer()
21   **until** nxtParent = tryParent
22   Connect to tryParent

There are many ways to find a new parent broker for B depending on the topology requirements. One is to look for an arbitrary broker which has less than b child brokers down the tree and use it as a new parent for a requesting broker. This way, the broker is integrated as a leaf into the tree and the degree of the broker topology can be maintained, however, it does not prevent the degeneration of the tree deterministically. An example for the procedure askForPlaceOrPointer() is depicted in Algorithm 15.

Algorithm 15 Example for procedure askForPlaceOrPointer() which maintains the broker degree b

---

23 **procedure** askForPlaceOrPointer()
24   **if** $|\mathcal{C}^{c^{\text{new}}}| < b$ **then**
25   **return** B
26   **else**
27   **return** random $C \in \mathcal{C}^{c^{\text{new}}}$
28   **endif**

The broker overlay is in a *correct state* if the parent and child broker relation between every broker in the system is consistent for the data structures colored with the values of c^{old} and c^{cur} at R (for the value of c^{new} at R the pointers may be inconsistent due to message propagation delays) and the tree which is defined by $\mathcal{P}^{c^{\text{old}}}$ and $\mathcal{C}^{c^{\text{old}}}$, and $\mathcal{P}^{c^{\text{cur}}}$ and $\mathcal{C}^{c^{\text{cur}}}$, respectively, is not partitioned (i.e., there is exactly one way from one broker to another). The value of $\mathcal{C}^{c^{\text{new}}}$ and $\mathcal{P}^{c^{\text{new}}}$ is treated differently as explained in the next section about reconfiguration. Partitions or cycles are detected as explained in the context of procedure onReceiveRec() (Algorithm 10).

Reconfiguration

The focus so far laid on the overlay network management and on how to handle control messages and notification routing. The attribute *color* has been introduced to synchronize actions on the overlay network layer with the publish/subscribe routing layer. All this was preparatory to incorporate reconfiguration into self-stabilizing publish/subscribe systems which is presented in this section.

Whenever a leaf broker wants to join or leave the overlay network or a link has to be replaced by another one, the topology of the broker network changes. When a reconfiguration should be implemented, the intended changes are sent to R, which collects them in the set $\mathfrak{R}$ and disseminates them in the next recolor message. Every broker that receives a recolor message carrying reconfiguration data that affects itself implements the change into its $\mathcal{P}^{c^{\text{new}}}$ and $\mathcal{C}^{c^{\text{new}}}$ pointers (procedure call

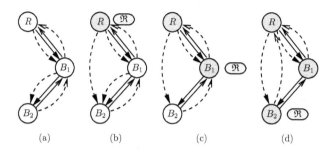

Figure 4.19: Example reconfiguration of the self-stabilizing publish/subscribe broker overlay topology

applyReconfig() on line 7 of procedure onReceiveRec(), Algorithm 10). The recolor message serves as a synchronizer to prevent race conditions when switching from one topology to another. Recolor messages are routed using $C^{c^{cur}}$ of every broker B which receives a recolor message (where the value of c^{cur} of B equals the value of c^{cur} of R). Thus, reconfigurations take two recolor messages to become active: one to disseminate the reconfiguration and one to activate it.

Figure 4.19 shows an example reconfiguration scenario (without faults), where B_2 shall be moved as a child broker from B_1 to R. The solid lines depict the parent/child relations for c^{cur} while the dashed lines depict the parent/child relation for c^{new} (Figure 4.19(a)). White brokers turn gray when they have received the recolor message and rotated their colors. The reconfiguration request is sent to R which incorporates it into its child broker pointers before sending it with the next recolor message in the set $\mathfrak{R}$ to B_1 over $C^{c^{cur}}$ (Figure 4.19(b)). On receiving the recolor message, B_1 updates its parent/child pointers $\mathcal{P}^{c^{new}}$ and $C^{c^{new}}$, i.e., R stays the parent of B_1 and B_1 will have no child brokers anymore (Figure 4.19(c)). Then, the recolor message is forwarded to B_2 which removes B_1 as the next parent broker and sets its $\mathcal{P}^{c^{new}}$ pointer to R (Figure 4.19(d)). The new parent/child pointers become active with the next recolor message disseminated by R.

As mentioned earlier, a change in the topology in general implies a change in the routing tables on the publish/subscribe routing layer. As the routing tables are regularly rebuilt from an initial routing configuration, the reconfiguration of the overlay topology can be incorporated by delaying the switch to the new topology in $\mathcal{P}^{c^{new}}$ and $\mathcal{C}^{c^{new}}$ long enough, such that they have been rebuilt completely when the switch to the new topology is executed. In the following, the self-stabilizing routing algorithm is described that is layered on top of the self-stabilizing broker topology.

Integration of Self-Stabilizing Routing

Recolor messages are used on the topology layer to trigger timeouts and coordinate reconfigurations. To achieve the latter, three topologies are held in form of colored parent/child pointers. On the routing layer, the color is used for two different purposes: (i) to rebuild the routing tables periodically and (ii) to avoid notification loss and duplicates.

It is necessary to periodically rebuild the routing tables since it is assumed that they can be perturbed arbitrarily. Therefore, the leasing mechanism described in Section 3.3 is used: clients regularly refresh their subscriptions and brokers use a second chance algorithm to remove stale entries from their routing tables. To incorporate reconfigurations into this mechanism, it is required that control messages (subscriptions and unsubscriptions) are colored with c^{new}, while notifications are colored with c^{cur}. Notifications and control messages are then forwarded and applied to the routing tables $T^{c^{cur}}$ and $T^{c^{new}}$, respectively. Thereby, it is ensured that notifications will be routed over the topology, the publishing broker belonged to at publication time. This way, duplicates are prevented, i.e., notifications sent multiple times to the same broker, which could only happen if a notification can be colored with multiple colors and takes different paths to the same broker in case of reconfigurations. The second chance algorithm is implemented through rotating the colors and initializing $T^{c^{new}}$ with a legal initial routing configuration init (Algorithm 9 on page 164, line 6, and Algorithm 10, line 6, respectively).

The algorithm has to take care that the routing table $T^{c^{new}}$ on every broker in the system is complete when the next recolor message is sent by R. Otherwise, notifications might get lost. Therefore, the refresh period ρ for every subscriber has to be chosen according to the values determined in Section 3.3:

$$\xi > \rho + 2 \cdot h \cdot \delta_{\max} \qquad (4.2)$$

A notification is colored with c^{cur} of the publishing broker. It may happen that a notification encounters a recolor message on its way to R. In this case, the following brokers already rotated their colors such that the color of the notification is now different from c^{cur} of those brokers. To be able to route the notification until it reached all intended receivers, the brokers also store the data structures colored with c^{old}, where c^{old} has the value of c^{cur} before the last recolor message has been received.

New (un)subscriptions are sent out immediately. Since a subscription may also encounter a recolor message on its way to R, control messages are colored with c^{new} of the issuing broker to avoid that subscriptions are present in $T^{c^{\text{old}}}$ and later $T^{c^{\text{new}}}$ but not in $T^{c^{\text{cur}}}$. Figure 4.20 depicts a situation, where a new subscription is issued shortly before a new recolor message is received. Before the subscription has reached every relevant broker, it encounters a recolor message which rotates the colors of the brokers. Due to the FIFO property of the communication channels, the color of the subscription equals the value of c^{cur} of every subsequently reached broker, because it "follows" the recolor message afterwards. At time t_r, when the subscription and the recolor message have reached every broker in the system, the new subscription is consistently incorporated into the routing tables of every affected broker. Due to the choice of the timeout ρ, the new subscription will also be incorporated into the routing table of every broker when the next recolor message is sent by R.

If the color of the subscription would have been the value of c^{cur} of B in this example, the subscription would have been incorporated into $T^{c^{\text{old}}}$ of every broker in the system from time t_r until time $t_0 + 2 \cdot \xi$. In this period, the subscription would not be present in $T^{c^{\text{cur}}}$ of the brokers such that "old" notifications would be

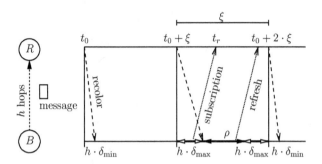

Figure 4.20: A new subscription encounters a recolor message on its way to R routed to B but not the current ones. Thus, notifications could be lost.

Self-Stabilization

A self-stabilizing system is guaranteed to eventually reach a legitimate state in case no fault occurs for a long enough time period. In this section, the different states the systems can end up in due to faults are discussed. The state is defined by the contents of the variables stored in RAM a broker uses. Those comprise the following values a broker holds:

(1) The values of its color variables

(2) Its child and parent pointers (overlay network layer)

(3) Its routing table entries

A corruption of these values may lead to the following faults:

(1) On the overlay network layer:

    (1a) Network partitioning

    (1b) Cycles

(2) On the routing layer:

(2a) Messages not forwarded or delivered although they should be

(2b) Messages forwarded or delivered although they should not be

In addition to that, brokers and links may crash and come up again due to transient faults. Since it is assumed that broker crashes and link failures are only transient, it is reasonable to concentrate on the time after the transient faults, when the brokers or links are up again—with a possibly corrupted state. Messages inserted due to faults can be modeled as faults that manifest in the state of the broker or as messages delivered although they have never been published. However, message inserted due to faults may be forwarded for at most time $h \cdot \delta_{\max}$ until they have left the system. In the following, all faults are discussed which may happen and how the system stabilizes itself afterwards.

Network Partitioning. If the network becomes partitioned, for example, because the parent and child broker pointers of two brokers are perturbed accordingly, the brokers in the part of the network which does not include R will eventually run into a timeout because they will not receive recolor messages any more. This case is handled in Algorithm 9, where a regular broker that runs into a timeout tries to rejoin the tree by contacting R.

Cycles. Cycles in the broker overlay topology may result from perturbed parent/child broker pointers, for example, if an ancestor B' of broker B is at the same time a child of B. In this case, B will send a recolor message to B' who has already received it. Due to the checks done in Algorithm 10, B' will not accept the recolor message. Accordingly, it will also not reply with an acknowledge message. Thus, B will remove B' eventually from his child brokers set (Algorithm 11 and Algorithm 12).

Cycles in the topology, where all parent/child broker pointers are consistent, are handled as network partitions because this case can only happen if the network is partitioned and the root broker of the partition without R additionally has a parent broker.

Perturbed Routing Tables. Regarding the routing tables, the same approach as presented in Chapter 3 is followed which relies on a precautionary reset. This guarantees that routing table entries which have been modified or inserted due to a fault will vanish eventually and that routing table entries that have been removed due to a fault will be inserted again.

Colors. The colors of a broker are stored in the variables c^{new}, c^{cur}, and c^{old} and can be perturbed arbitrarily. However, every recolor message rotates all color values and sets the value of c^{new} to the color stored in the recolor message (cf., Algorithm 10). Thus, the colors are eventually consistent with those of R.

Root Broker. The root broker R plays a central role in this algorithm because it functions as the synchronizer as well as the central contact which is responsible for coordinating reconfigurations. For the whole system being self-stabilizing it is therefore necessary to implement R in a self-stabilizing fashion. Although R is already self-stabilizing according to the timeout mechanism, it can be implemented more robust by using a root group. Such a *root group* consists of several brokers that take over R's task in a predefined, globally known order in case the previous root broker fails. Details on this topic can be found in [60].

Stabilization Time

Every corrupted routing table entry will be removed after at most

$$3 \cdot \xi + h \cdot (\delta_{\text{max}} - \delta_{\text{min}})$$

Assume that in the worst case the table colored with c^{new} is corrupted by inserting a bogus entry right after a recolor message has been received. In this case it will take at most three recolor messages until the wrong routing table entry has vanished from $T^{c^{\text{old}}}$. A recolor message is sent every ξ, and will take at least $h \cdot \delta_{\text{min}}$ and at most $h \cdot \delta_{\text{max}}$ until it has reached the last broker. Thus, a recolor message may arrive at most $h \cdot (\delta_{\max} - \delta_{\min})$ after the last recolor message received before.

(Un)subscriptions are colored with c^{new} of the issuing broker. If a bogus unsubscription message $\neg s$ is induced into the system, it may arrive in the worst case right after a respective subscription s arrived at a broker, but less than $\rho + h \cdot (\delta_{\max} - \delta_{\min})$ before the next recolor message arrives such that the unsubscription is not overridden by a refreshing subscription before the next recolor message arrives. Thus, the routing table misses the subscription in the following which is active until the color of the broker turns the value of c^{new} to the value of the corrupted routing table again and initializes it subsequently. This takes no more than two recolor messages or time $2 \cdot \xi + h \cdot (\delta_{\max} - \delta_{\min})$. In this case is takes at most time

$$2 \cdot \xi + 2 \cdot h \cdot (\delta_{\max} - \delta_{\min}) + \rho$$

until the system is stabilized again.

A partitioning of the network forces brokers to rejoin the tree. It takes at most $\xi + h \cdot \delta_{\max}$ (cf. Equation 4.1) until a broker B recognizes a partitioning. The time needed for re-integrating B into the overlay network depends on the algorithm used. It is reasonable to assume that finding a parent broker for B takes no more than time ξ. It takes at most another time ξ until a recolor message is sent out by the root broker that carries the reconfiguration which integrates B into the system. Finally, it takes at most $\xi + h \cdot (\delta_{\max} - \delta_{\min})$ until every broker has incorporated the new configuration into its state and the new configuration is active, i.e., the new configuration is implemented in the variables colored with c^{cur} of R. Thus, the system is stabilized after at most

$$\xi + h \cdot \delta_{\max} + 2 \cdot \xi + \xi + h \cdot (\delta_{\max} - \delta_{\min}) = 4 \cdot \xi + h \cdot (2 \cdot \delta_{\max} - \delta_{\min})$$

Thus, the stabilization time Δ_r of the self-stabilizing algorithm is given by

$$\Delta_r = 4 \cdot \xi + h \cdot (2 \cdot \delta_{\max} - \delta_{\min}) \tag{4.3}$$

4.5.4 Ordering

In contrast to the advanced coordinated reconfiguration algorithm, the coordinated reconfiguration algorithm for layered self-stabilizing publish/subscribe systems does not respect notification ordering yet. However, it is also possible to incorporate FIFO-publisher and causal ordering into this algorithm. In the following, two extensions are sketched that are able to provide notification ordering guarantees in the face of reconfigurations.

Causal Ordering. In general, it is save to queue all notifications for one recolor period after a reconfiguration has been implemented. Doing this, it is guaranteed that all notifications that have been published in the old topology have reached their destinations when disseminating the notifications in the new topology after the reconfiguration. This way, it is possible to provide causal ordering guarantees.

FIFO-Publisher Ordering. The FIFO-publisher ordering of notifications can only be disturbed if a notification passes a link that was part of a reconfiguration, i.e., which has been added to the topology. Thus, a more optimized solution can be provided than that presented for causal ordering. Here, it suffices to delay notifications for one recolor period which are going to be sent over a link that has been added to the topology due to a reconfiguration. Such a link can be detected by simply comparing the neighbor broker pointers colored with c^{cur} and c^{old}. This way, reconfigurations do not affect all notifications disseminated but only those which are directly affected by it.

Both mechanisms can be easily integrated into the algorithm but induce a notification delay just like the advanced coordinated reconfiguration algorithm. Another option would be to follow the approach presented by Zhao et al. [165] which applies vector clocks. This approach requires more effort for integration but may lower the average notification delay due to reconfigurations.

4.5.5 Discussion

The coordinated reconfiguration algorithm presented for layered self-stabilizing publish/subscribe systems is the first algorithm for self-stabilizing publish/subscribe systems that allows for reconfiguration at runtime without message loss. Using this algorithm, it is possible to incorporate reconfigurations at runtime without service interruption.

Advertisements

The algorithm discussed does not support advertisements. It is easy to support advertisement by following the same approach as discussed in Section 3.4.2. Therefore, the value of ξ should be increased such that there is enough time to (i) disseminate the recolor message to all brokers and incorporate necessary reconfigurations, (ii) to build the subscription routing tables, and (iii) to build the notification routing tables accordingly.

Subscription Activation Delay

New subscriptions are colored with c^{new} of the issuing broker. This may increase the delay for a subscription to become active. An option to reduce the delay would be to send new subscriptions additionally using the topology colored with c^{cur} of the issuing broker. The price would be an increased message overhead.

Parallel Reconfigurations

Parallel reconfiguration requests can be handled by R. If they are interfering or contradicting they can be rejected. If it is obvious that they can be executed in parallel, they are disseminated. Otherwise, they need to be serialized. However, R may not be able to oversee the consequences of every combination of reconfigurations. If reconfigurations are sent to R which lead to an incorrect topology, this may lead to a fault from which the system will recover eventually.

Scalability

In large-scale heterogeneous scenarios, the algorithm proposed may lead to long stabilization times and long average delays for the implementation of reconfigurations. Here, it is beneficial to not use one big dissemination tree but a tree which consists of several subtrees in a hierarchical fashion: the subtrees are viewed as leaves represented by the root of the subtree in the main tree and all trees are self-stabilizing using the proposed algorithms. This way, longer communication delays over wide-area connections would have less impact on the stabilization time and reconfiguration delays in subtrees of the system. However, this approach restricts reconfigurations to subtrees which is sensible in many scenarios, where administrative domains do not overlap.

4.6 Related Work

Most of the relevant related work has already been discussed in Section 4.4.1 and Section 4.5.1. In this section, related work not covered yet is discussed.

Self-Stabilization. The possibility of exploiting self-stabilization to create systems that are able to manage themselves to a certain degree has recently gained interest in the growing research communities that work on ad hoc and sensor networks [74]. In these scenarios, (huge) numbers of nodes are deployed without any infrastructure like central access points or routers and it is expensive (if not impossible sometimes) to manage them manually. These scenarios are subject to dynamic changes that may lead to changes of the topology and it cannot be expected that all reconfigurations of the topology do not hurt the safety properties of the system.

Kakugawa and Yamashita introduce the class of *dynamic reconfiguration tolerant* (DRT) algorithms, where a system that is in a legitimate state is guaranteed to not leave this state given that no faults or certain network reconfigurations occur [90]. The goals of their work are very similar to those of superstabilizing protocols with the difference that the passage predicate that must hold during reconfiguration of a superstabilizing system equals the correctness predicate of the self-stabilizing system itself [73]. Along with their definition, the coordinated reconfiguration algorithm for self-stabilizing publish/subscribe systems belongs to the class of DRT self-stabilizing algorithms. However, the authors do not generalize their approach beyond the example given for a DRT self-stabilizing token reconfiguration algorithm.

Message Ordering. The advanced coordinated reconfiguration algorithm is able to guarantee FIFO-publisher or causal ordering of delivered events. Ordering is also an issue for systems with mobile clients, where notification ordering may also get mixed up due to wandering clients (and implicit reconfigurations that are triggered thereby). Lwin et al. discuss the case of causal ordering in mobile environments assuming a simplified model, where clients are removed and added only as leafs and no brokers join and leave the system [106]. They use a vector clock like mechanism to ensure a FIFO-publisher ordering for (un)subscriptions and notifications and use queues to delay messages. The authors do not discuss the case, where the broker network is reconfigured.

A distributed approach is taken by Lumezanu et al. by subdividing the task of message ordering into sequencing atoms which assign sequence numbers to messages [105]. Their approach does not rely on one acyclic broker overlay topology with FIFO channels for notification delivery (which guarantees causal ordering given that the brokers process messages in FIFO-publisher order) as done here. Instead, the notion of groups of subscribers is introduced together with dedicated mechanisms for message ordering between different groups of subscribers.

4.7 Discussion

In this chapter, the general problem of reconfiguring the broker overlay topology of a publish/subscribe system at runtime without service interruption is discussed and analyzed. The notion of reconfiguration used implies that reconfigurations can be delayed for a finite time period. Moreover, they transform the broker overlay topology from one legal topology to another one. This marks an important difference between a fault and a reconfiguration. The main challenges in implementing reconfigurations are identified and analyzed and new algorithms are presented which are the first to support the reconfiguration of acyclic broker topologies for conventional and self-stabilizing publish/subscribe systems without message loss while keeping message orderings, thus enabling seamless reconfigurations at runtime.

The advanced coordinated reconfiguration algorithm for conventional broker overlay networks presented is able to maintain FIFO-publisher and causal ordering of notifications. Besides that, it always ensures FIFO-publisher ordering for (un)subscriptions which is required to maintain the correctness of a publish/subscribe system in face of reconfigurations. The algorithm relies on the coloring of messages and additional queues for (un)subscriptions and notifications (if an ordering of notifications is required). It is the first algorithm that realizes the reconfiguration of an acyclic publish/subscribe broker overlay topology at runtime without message loss. Even more, it is able to provide ordering guarantees for messages no other reconfiguration algorithm supports yet. The integration into the general publish/subscribe model showed that the required changes are reasonably small and, thus, feasible in a wide range of publish/subscribe systems. Furthermore, it was shown in a simulation study that the new algorithm outperforms the naive STRAWMAN approach with respect to message overhead and delay—although the new approach relies on additional messages that are needed for coordination purposes. The advanced coordinated reconfiguration algorithm requires a fault-free scenario and is, thus, an important supplement to the algorithms presented in related work which are designed for error-prone environments, where reconfigurations are triggered by unpredictable link failures.

Reconfiguring arbitrary self-stabilizing systems with respect to certain guarantees implies some subtle problems and requires further research efforts. These problems were analyzed with the outcome that simply layering self-stabilizing algorithms on the topology and the routing layer is not feasible in publish/subscribe systems because the contents of the routing tables are depending on the broker overlay topology. The new approach presented here relies on a combination of self-stabilizing content-based routing as introduced in Chapter 3 with a self-stabilizing broker overlay network that accepts arbitrary acyclic broker topologies in case no fault occurs. In order to meet the dependencies between both layers, a coloring mechanism was introduced that coordinates actions on them. Thus, it was possible to prevent message loss during reconfiguration and guarantee causal and FIFO-publisher ordering. The coloring mechanism described is not limited to self-stabilizing publish/subscribe systems. It is a general principle which can be used to realize seamless reconfigurations in other layered self-stabilizing systems as well. However, the new approach currently requires a dedicated root broker which is responsible of coordinating reconfigurations. It may be sensible to realize this broker in form of a cluster to prevent a bottleneck or to take a modular approach for an improved scalability. Both issues require more research and remain open for future work.

The algorithms presented in this chapter are a necessary prerequisite to combine the (self-stabilizing) routing layer with an adaptive reconfiguration mechanism which runs on top of the publish/subscribe layer and issues reconfiguration stimuli. These reconfigurations can be implemented by the lower layers as described in this chapter. Thus, the algorithms are building blocks for self-managing notification services in general. Using them, it is now possible to manage publish/subscribe systems at runtime without service interruption which improves the applicability in dynamic environments that are often faced in practice.

5 Self-Optimizing Notification Service

Contents

5.1 Introduction

Many applications that build on the publish/subscribe communication paradigm like multiplayer online games [21], the dissemination of news articles [125, 135], security network monitoring [65], and online auctions [26, 103] are subject to dynamic behavior of the system's publishers and subscribers. For example, the interest in certain news topics may vary in time and region as well as the number of notifications published that belong to these topics. This behavior may change arbitrarily due to the free will of the users. Moreover, not only the message flows change over time. The processors and the communication links are shared by various applications such that the properties of the communication links as well as the processing capabilities available for serving the publish/subscribe system on the individual brokers may change over time. Thus, according to the performance function applied, a broker overlay topology that takes care of message forwarding may be "better" than another one or not—depending on the current setting.

In this chapter, an approach is presented to enable publish/subscribe systems to autonomously optimize their broker overlay topologies according to a performance model which takes processing and communication costs into account. The goal is to render the broker topology of a publish/subscribe system self-optimizing, i.e., to enable it to adapt to dynamic changes without human intervention in order to increase its performance. This ability is of great value in large and complex systems, where human administrators are not able to manually administer the system anymore. It is also useful for systems, where no administrator is available but a good performance is still required or useful, for example, in small-scale e-home scenarios. Besides automated fault management as described in Chapter 3, self-optimization is one key component of self-managing publish/subscribe systems.

In the next section, the system model used is introduced together with the cost model used. Then, a formal specification of the problem statement is given in Section 5.3. It is shown that the problem of finding an optimal broker overlay network topology for a given distribution of clients is NP-complete and conclude that it is sensible to use a heuristic to approach this problem. Two heuristics that have been proposed in literature to approximate a solution for the optimal broker overlay network topology problem are discussed in Section 5.4. It is shown that they both rely on assumptions that do not hold in general. Accordingly, a new heuristic is proposed in Section 5.5 together with a framework that can also be used to easily integrate other heuristics into publish/subscribe systems that rely on acyclic broker overlay topologies. With the new heuristic, brokers have a limited view on the network and try to optimize the topology of their neighborhood according to common traffic as well as communication and processing costs. The goal is that the local optimizations carried out by each broker emerge in a good (in terms of costs) overall broker topology. To evaluate the performance of the new heuristic, it is compared to the two exemplary heuristics in Section 5.6. The simulation-based evaluation comprises various aspects like the total performance, the ability to adapt to changes, the resulting topology characteristics, and the flexibility with respect to the distribution of cost.

5.2 System Model

In the system model used, the notification service that takes care of forwarding notifications from publishers to interested subscribers consists of a set of brokers $\mathcal{V}$. A connected network graph $G = (\mathcal{V}, \mathcal{E})$ is the starting point. This graph consists of the brokers $\mathcal{V}$ and the potential overlay network links $\mathcal{E}$. Usually, each broker $v \in \mathcal{V}$ can potentially connect to every other broker in $\mathcal{V}$. However, some edges might be omitted, for example, due to administrative reasons.

All notifications N published are distributed using a single *spanning tree* T of G, where each $n \in N$ is published by a dedicated broker $P(n) \in \mathcal{V}$ and must be delivered to a respective set of brokers $\mathcal{S}(n) \subseteq \mathcal{V}$. In the following, T is called the *broker overlay (network)*. An example for a broker overlay network that is embedded in a general graph is shown in Figure 5.1.

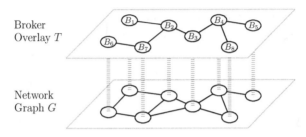

Broker
Overlay T

Network
Graph G

Figure 5.1: Example graph with an embedded broker overlay

5.2.1 Cost Model

To measure the performance of a publish/subscribe system, a cost model is introduced that comprises processing costs and communication costs. For every link $e \in \mathcal{E}$, the *communication costs* c_e are defined. Accordingly, for every broker $v \in \mathcal{V}$, the *processing costs* p_v are defined. The communication costs occur when a notification is sent over a link, while the processing costs occur when a notification is processed by a broker.

The cost metric is abstract and can comprise arbitrary aspects depending on the application or the desired goals. The communication costs may correlate, for example, with the delay or the bandwidth of a link or a combination of both. The processing costs may correlate with the processor speed of a broker its main memory or, again, a combination of both, for example. Every message that is forwarded in the broker overlay network causes costs at each broker and link it passes. In the following, it is assumed that those costs change with a moderate speed, because it would otherwise be hard to adapt the topology in time. Analogous, it is not the aim to be able to adapt to sudden load spikes which would require an extremely fast adaptation mechanism. Moreover, this would require sophisticated application specific prediction mechanisms because a sudden load spike may need to be predicted in advance in order to be able to react adequately.

The cost model is used in the following to measure the "performance" of a broker overlay network topology with respect to a given client behavior. The model is simple but very powerful because both metrics can be a combination of various other metrics and be normalized and combined arbitrarily, being able to capture a wide range of optimization targets for the broker overlay topology of a publish/subscribe system. The cost model allows judging the quality of a broker topology with respect to the underlying physical network topology.

In contrast to other approaches, the model takes processing costs into account. This is sensible because advanced matching algorithms, e.g., for XML documents, need non-negligible periods of time for drawing routing decisions for incoming messages [7, 36].

The cost model implicitly assumes that the processing costs of messages do not vary too much for different messages because it is assumed that the processing costs do not depend on the individual message. Moreover, it assumes that the number of neighbors of a broker does not have a significant influence on the processing costs. Both assumptions may not hold in general. However, it is reasonable to believe that working with average values for the processing costs per message is sufficient for many scenarios.

5.3 Problem Statement

The goal here is to improve the performance of a publish/subscribe system by adapting the structure of the broker overlay network. To measure the performance, communication and processing costs are considered which accumulate per message.

In static environments, where client behavior and network structure do not vary over time, it is sufficient to construct the broker overlay network only once in order to optimize the performance of the whole system.

In the following, this static publish/subscribe overlay optimization problem is considered first in the introduced system model and it is shown that it is NP-hard, even if the notification flows are known in advance. For large-scale settings, this means that heuristics must be used to find an acceptable solution in a reasonable time for this problem. Then, it is argued that the problem is even harder in dynamic environments. To keep the system within good working conditions, it is necessary to additionally *adapt* the system to *unpredictable changes* at runtime.

5.3.1 Basics

Figure 5.1 depicts an example of a broker overlay network in a graph consisting of eight brokers. Instead of using the whole spanning tree for the distribution of a notification n, only the minimal connected subtree of T is used which contains the publisher and the subscribers of n. This subtree is called the *delivery tree* of n and is defined as follows.

Definition 10 (Delivery Tree). *For a given broker overlay tree T and a notification n the delivery tree D_T^n of n with respect to T is defined by*

$$D_T^n := (\mathcal{V}_T^n, \mathcal{E}_T^n) \tag{5.1}$$

where $\mathcal{V}_T^n$ is the set of brokers and $\mathcal{E}_T^n$ is the set of overlay links contained in the minimal subtree of T that connects the brokers hosting the subscribers $\mathcal{S}(n)$ and the publisher $P(n)$.

According to the definition of the delivery tree and since n is sent over all links in $\mathcal{E}_T^n$ and is processed by all brokers in $\mathcal{V}_T^n$, the cost for distributing n is given by

$$\mathrm{cost}(n, T) := \sum_{\forall v \in \mathcal{V}_T^n} p_v + \sum_{\forall e \in \mathcal{E}_T^n} c_e \tag{5.2}$$

The delivery tree of $n \in N$ is determined by the set of brokers which consists of the brokers hosting the publisher and the subscribers of n. This set is defined as $V(n) := \{P(n)\} \cup \mathcal{S}(n)$. $\mathcal{V}_T^n$ and $V(n)$ can differ significantly. Assume that in the example broker overlay depicted in Figure 5.1, broker B_2 publishes a notification n in which only B_5 is interested. In this case, $\mathcal{V}_T^n = \{B_2, B_3, B_4, B_5\}$ includes B_3 and B_4 even though they are not interested in n because $V(n) = \{B_2, B_5\}$.

Obviously, the total cost of forwarding one specific notification in the broker overlay strongly depends on its delivery tree and, thus, on the spanning tree. The overall cost for distributing all notifications in N using one spanning tree T is then given by

$$\text{cost}(N, T) := \sum_{n \in N} \text{cost}(n, T) \qquad (5.3)$$

5.3.2 Static Case

In the model used here, a single spanning tree T is used for distributing all notifications. To optimize the costs it is, thus, necessary look for the spanning tree which minimizes $\text{cost}(N, T)$. Now the static optimization problem is formally defined.

Definition 11. (Publish/Subscribe Overlay Optimization Problem (PSOOP)) *Given a graph $G = (\mathcal{V}, \mathcal{E})$, communication costs c_e for all $e \in \mathcal{E}$, processing costs p_v for all $v \in \mathcal{V}$, a set of notifications N, and a function $V : N \to 2^{\mathcal{V}}$, determine the spanning tree T of G which minimizes $\text{cost}(N, T)$.*

Complexity

In the following, it is shown that the decision problem which corresponds to the PSOOP is NP-complete. Then, it is proved that PSOOP is NP-hard and that it is sensible to apply heuristics to solve the problem.

Definition 12. (Publish/Subscribe Overlay Decision Problem (PSODP)) *Given a graph $G = (\mathcal{V}, \mathcal{E})$, communication costs c_e for all $e \in \mathcal{E}$, processing costs p_v for all $v \in \mathcal{V}$, a set of notifications N, a function $V : N \to 2^{\mathcal{V}}$, and a bound $C_{\max} \in \mathbb{N}^+$, is there a spanning tree T of G with $\text{cost}(N, T) \le C_{\max}$?*

PSODP is a generalization of the *optimum communication spanning tree* (OCST) problem [77] which deals with the efficient construction of telecommunication networks interconnecting cities. In the OCST problem, each link $e \in \mathcal{E}$ of the graph $G = (\mathcal{V}, \mathcal{E})$ is labeled with communication costs w_e (e.g., with the length of the connecting wire). Additionally, the communication requirements $r(\{u, v\})$

between any two vertices $u, v \in \mathcal{V}$ are known. The problem consists of finding a spanning tree T for G with costs that lie below an upper cost bound $C_{\max}$.

Let $W(\{u, v\})$ be the function which returns the sum of the link costs of those links that make up the path connecting u and v in the spanning tree T. Then, the overall costs of a spanning tree T are

$$\text{cost}(T) := \sum_{u,v \in \mathcal{V}_T} r(\{u, v\}) \cdot W(\{u, v\}) \tag{5.4}$$

Definition 13. (OCST Decision Problem) *Given a graph $G = (\mathcal{V}, \mathcal{E})$, communication costs c_e for all $e \in \mathcal{E}$, communication requirements $r : \mathcal{V} \times \mathcal{V} \rightarrow \mathbb{N}^+$, a bound $C_{\max} \in \mathbb{N}^+$, is there a spanning tree T of G with $\text{cost}(T) \leq C_{\max}$?*

The OCST decision problem has been proven to be NP-complete in 1978 [88]. It can be easily shown that an instance of the OCST decision problem can be transformed into an instance of the PSODP problem using restriction: set the processing costs of the nodes to 0 and consider only the communication costs. Thus, PSODP is NP-hard. PSODP is also in NP because a deterministic Turing machine can verify a guessed solution in polynomial time, i.e., compute the actual overall costs of an overlay. Thus, PSODP is NP-complete. Hence, in order to solve this problem efficiently, it is necessary to apply heuristics. A brute force search for an optimal spanning tree would result in trying all n^{n-2} spanning trees that can be created in a fully connected graph with n nodes according to Cayley's formula [40].

Approximation Algorithms

Please note that heuristics for approximating the OCST optimization problem in an efficient way already exist. However, they all rely on global knowledge. In [124], Peleg and Reshef propose an algorithm that approximates the optimal solution for the OCST problem in time $\mathcal{O}(\log_2 |\mathcal{V}|)$ with bound $\mathcal{O}(\log^3 n)$. However, to be able to apply the solution, $W(u, v)$ must be conform to the triangle inequality which is not necessarily true on the Internet.

The *minimum routing cost spanning tree* (MRCST) problem is a special case of the OCST problem, where all nodes u and v have the same communication requirements (i.e., $r(\{u, v\}) = 1$). For this case, a polynomial approximation scheme exists [158]. Two further special cases of the OCST problem that additionally consider vertex weights are the *product requirement optimum communication spanning tree* (PROCT) and the *sum-requirement optimum communication spanning tree* (SROCT) problem which can be solved with a 1.557- (PROCT) and 2-approximation (SROCT) in $\mathcal{O}(n^3)$ and $\mathcal{O}(n^5)$, respectively [157]. For the PROCT problem, the communication requirement between two nodes equals the product of the weights of both nodes, while for the SROCT problem, the communication requirement between two nodes equals the sum of their weights. Both models do not fit for the scenario assumed here because processing costs are not considered.

In [102], Li and Bouchebaba present a heuristic based on a genetic algorithm. This heuristic can lead to optimal or near-optimal solutions but requires global knowledge, too.

All approximation algorithms presented so far require global knowledge and are, thus, not suited for a large-scale distributed publish/subscribe system, where global knowledge is not feasible. Another severe limitation with using these approximation algorithms in a distributed publish/subscribe system is the anonymous communication style. It decouples publishers and subscribers and makes it hard to determine the mutual communication requirements between all brokers which are required for applying the approximation algorithms discussed. Moreover, the mutual communication requirements are not predetermined because they do not only depend on the interests of the brokers (in terms of subscriptions) but also on the notifications published.

5.3.3 Dynamic Case

In the previous section, the problem of finding an optimal broker overlay network has been formalized in a static setting, where all parameters are known (the notifications, their respective publishing/consuming brokers, as well as the processing and communication costs of brokers and links, respectively). However, in many scenarios this setting is not realistic due to a number of reasons. The set of notifications that brokers publish and subscribe for may vary over time. This affects the number of notifications to be forwarded and the parts of the overlay network through which they flow. The network topology may also change including processing as well as communication costs.

The above changes can, to a great extent, affect not only the operating costs at any point in time, but also the accumulated overall costs. In these cases, better results (i.e., lower costs), can be achieved by adapting the overlay network whenever a significant advantage may be gained. Since the changes are in general not known in advance, a typical on-line problem is faced. Thus, in order to adapt, it is necessary to rely on gathered data about the past for predicting the future. The approach presented constantly gathers data about notifications published and consumed as well as about processing and communication costs in order to derive potential adaptations that lower the operating costs.

Since the presented solution is targeted at large-scale systems, an algorithm based on global knowledge is not feasible. The new heuristic presented in Section 5.5 thus adapts the overlay network based on local knowledge only. It assumes that the overall setting changes with a moderate speed such that sophisticated (and application dependent) prediction mechanisms are not required.

5.4 Example Heuristics

The previous section showed that the problem of finding an optimal broker overlay topology for a dynamic publish/subscribe system is hard. In this section, two heuristics are presented that try to optimize the broker overlay topology. They rely on greedy algorithms and global knowledge and have both their individual drawbacks since they do not consider processing costs. Decentralized versions based on local knowledge will serve as a benchmark later in the evaluation section.

5.4.1 Minimum Spanning Tree

The *minimum spanning tree* (MST) heuristic performs best if the consumption of notifications is uniformly distributed between all brokers and the processing costs of all brokers are equal. In this case, a minimum spanning tree with respect to the communication costs is a good choice for the broker overlay network topology. The problem of building a minimum spanning tree is well researched and is, thus, not further discussed here [96, 131].

5.4.2 Maximum Associativity Tree

The *maximum associativity tree heuristic* or *maximum interest tree* (MIT) heuristic is based on one of the first approaches targeted at the optimization of publish/subscribe broker overlay topologies and has been published by Baldoni et al. [15, 16]. It focuses only on the notifications consumed by brokers and leaves network metrics aside. It is discussed in the following before the maximum interest tree heuristic is introduced.

Associativity Metric Based on Subscriptions

The original approach by Baldoni et al. explicitly concentrates on tuning the performance of the notification service by exploiting structural similarities of brokers and clustering them accordingly. The authors propose a distributed algorithm which considers the interest of each broker and builds an *associativity* metric from the intersection of these interests. Based on this metric, the algorithm tries to connect brokers with a high mutual associativity value to increase the overall associativity of the system. Thereby, the algorithm aims at decreasing the latency of notifications by reducing the average number of hops they travel through the system.

The authors derive a broker's *zone of interest* from the size of the notification space covered by the broker's local subscriptions [16]. Assuming that notifications are distributed uniformly, larger overlapping zones of interest result in a greater number of identical notifications being consumed by the brokers. To avoid extremely degenerated topologies a very limited network awareness is added by simply introducing a manually chosen upper bound for the costs a new overlay link can have. On this level of abstraction, it is not taken into account that even brokers exhibiting a low associativity value can be successfully deployed to decrease the network traffic and, thus, increase the system's performance.

The implicit assumption for using the associativity metric which is based on the comparison of filters is that notifications are distributed uniformly in the notification space. In this case, larger overlapping zones of interest result in a greater number of identical notifications consumed. However, this limits the applicability of the algorithm. Apart from this, the computation of the associativity is quite costly and difficult. This is why the authors proposed another application layer metric to derive the associativity value from.

The basic algorithm works as follows. It is triggered when a broker B_i believes that it is possible to increase the associativity of the system, i.e., when it seems likely that there is another broker behind a neighbor broker B_j that shares more interest with B_i than B_j does. To reason about this, every broker needs to know the local routing entries of every neighboring broker (which is not discussed in greater detail in the papers). If B_i manages a significantly bigger zone of interest in its routing table for B_j than B_j does for its local clients, a request message is sent that tries to find a broker that has a bigger associativity with B_i than B_j has. If such a broker B_k is found, a link on the path from B_i to B_k has to be removed to prevent a cycle when B_i and B_k are directly connected. The link connecting two brokers with the minimum associativity is chosen to be removed, if this associativity is smaller than the associativity between B_i and B_k.

Associativity Metric Based on Common Traffic

To circumvent the problem of determining overlapping zones of interests using filters, the authors introduce a history of the local events consumed or published by each broker to calculate the associativity based on the intersection of the set of messages consumed by two brokers [15]. Thereby, they try to become independent of the distribution of notifications and consider only the actual notification flow, which is similar to the basic approach taken here. In recent work, the authors propose the use of a special request message which carries the history of a broker. This message is sent out regularly and is evaluated by all brokers it reaches in order to determine the associativity of the receiving broker with the sender [17]. Finally, the metric focuses solely on the application layer and does not take processing load and costs for links on lower layers into account (apart from limiting the costs of new overlay links as discussed above). This is due to the main goal of the authors to minimize the number of pure forwarding brokers (i.e., brokers that only forward messages and do not have any local clients that issue subscriptions). It is, thus, still not taken into account that forwarders can be successfully used to decrease network traffic and, thus, increase system performance.

Maximum Interest Tree Heuristic

The example heuristic that creates maximum associativity trees assumes that it knows the set of notifications consumed by every broker in the system. Then, it reconfigures the network until all brokers with common interest are "close" to each other. Therefore, the link between every pair of brokers is labeled with the number of notifications they share. The maximum associativity tree is then built by using a modified minimum spanning tree algorithm such that a maximum spanning tree is built with respect to the labels of the links.

In the following, the terms "common traffic" and "common interest" of a set of brokers is used, when talking about identical notifications consumed by all brokers in this set.

5.4.3 Disadvantages

As already mentioned above, both heuristics rely on implicit assumptions regarding the distribution of notifications in the notification space and the processing costs of the brokers. In this section, an example case is shown, where both heuristics do not manage to find an optimal spanning tree for notification dissemination in a simple broker overlay network topology. This example motivates the heuristic presented in the following section.

Assume that the broker overlay network consists of three brokers B_1, B_2, and B_3 as depicted in Figure 5.2(a) on the facing page. The link labels depict the communication costs and the node labels depict the processing costs of each broker. Broker B_2 has a client that publishes notifications for which clients at B_2 and B_3 subscribed. Of ten notifications published by the publisher at B_1, B_2's client is interested in five notifications, while B_3's client is interested in all ten notifications published at B_1, i.e., B_1 and B_3 share a common traffic of $I_{1,3} = 10$. In addition to that, a publisher at B_1 publishes one notification for which only one client at B_2 is subscribed. Thus, B_1 and B_3 share a common traffic (or interest) of 6 notifications ($I_{1,2} = 6$).

Minimum Spanning Tree. Using the MST heuristic the topology depicted in Figure 5.2(b) on the next page is created because link $\overline{B_1 B_3}$ is the most expensive one. However, the total cost of this topology is determined not only by the communication costs of the links but also by the processing costs of the brokers. Thus, the cost of the topology under the given traffic sums up to

$$
\begin{aligned}
\text{cost(MST)} \quad &= \quad 10 \cdot (p_2 + c_{1,2} + p_1 + c_{1,3} + p_3) \\
&\quad +1 \cdot (p_1 + c_{1,2} + p_2) = 129
\end{aligned} \tag{5.5}
$$

Maximum Interest Tree. The MIT heuristic only considers the common traffic shared by brokers. Thus, the topology that results from applying the MIT heuristic looks like the one depicted in Figure 5.2(c) on the facing page. The cost of this

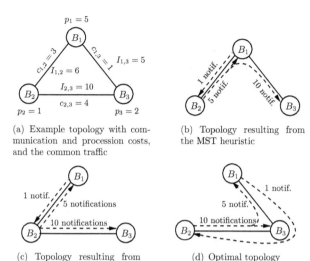

(a) Example topology with communication and procession costs, and the common traffic

(b) Topology resulting from the MST heuristic

(c) Topology resulting from the MIT heuristic

(d) Optimal topology

Figure 5.2: Example topology showing topologies created by the MIT and MST heuristic (B_1 is not interested in any of 10 notifications published by B_2 and consumed by B_3)

topology under the given traffic sums up to

$$\begin{aligned} \text{cost(MIT)} \;=\; & 10 \cdot p_2 + 10 \cdot (c_{2,3} + p_3) + 5 \cdot (c_{1,2} + p_1) \\ & + 1 \cdot (p_1 + c_{1,2} + p_2) = 119 \end{aligned} \tag{5.6}$$

Optimal Topology. From the calculations carried out above it is obvious that the topology created by the MIT heuristic is cheaper than the one created by the MST heuristic. However, it is not the cheapest topology possible which is depicted in Figure 5.2(d) and causes the following costs:

$$\begin{aligned} \text{cost(opt)} \;=\; & 10 \cdot (p_2 + c_{2,3} + p_3) + 5 \cdot (c_{1,3} + p_1) \\ & + 1 \cdot (p_1 + c_{1,3} + p_3 + c_{2,3} + p_2) = 113 \end{aligned} \tag{5.7}$$

Conclusions. The example shows the possible disadvantages of using both heuristics. However, both are an important benchmark for the solution presented in the next section. A minimum spanning tree is for example used when implementing publish/subscribe communication over multicast [120] and the MIT heuristic is related to the first approach published to enable broker overlay networks to optimize themselves.

5.5 Cost and Interest Heuristic

The previous sections showed that the problem of optimizing the broker overlay topology in a publish/subscribe system regarding communication and processing costs is hard. Existing solutions have several drawbacks and are expected to only provide a good solution for special cases (e.g., if the processing costs are negligible or if global knowledge is available). Approximation algorithms exist for the OCST problem, however, they do not consider processing costs. In this section, a new approach is presented to tackle the PSOOP which is based on a heuristic. It is decentralized such that it does not require global knowledge. Section 5.5.1 explains the basic idea and basic mechanisms used. Section 5.5.2 and 5.5.3 explain two important phases of the algorithm followed by a description of how to integrate the algorithm into the publish/subscribe broker model introduced in Chapter 3, Section 5.5.4 (cf. Figure 4.9 on page 132).

5.5.1 Basic Idea

In reasonably large systems it is not realistic to assume that global knowledge is feasible. Thus, a decentralized approach is followed that is based on local knowledge only. It follows the principle of local optimizations which are expected to *emerge* in a "good" global structure with respect to the performance of the system in whole. Therefore, each broker in the system gains knowledge about a reasonably large set of brokers in its neighborhood and optimizes this neighborhood by reconfiguring the topology it overlooks.

To provide the brokers with knowledge about their neighborhood, every broker B regularly broadcasts a message m_B^K to all brokers in its neighborhood $\mathcal{N}_\eta(B)$. This neighborhood consists of all brokers that are no more than η hops away from B in the broker overlay network. To not overload the network, it is sensible to use only small values for η. The information contained in m_B^K is then used by every receiving broker to obtain and update knowledge about its neighborhood and B in particular.

The size of m^K should be kept small in order to not overburden the network and, thus, decrease the performance of the publish/subscribe system. Similarly, the time period between sending two subsequent broadcast messages by a broker should be kept reasonably small.

To estimate the common traffic of two brokers (i.e., the notifications both consume or publish), a cache is introduced on every broker. Using this cache, it is possible to track the notifications a broker delivers to or receives from its local clients. By comparing their caches, brokers can, hence, reason about the amount of common traffic they share. The contents of the cache is then included in m_B^K broadcasted by B. Based on this information, the brokers in $\mathcal{N}_\eta(B)$ are able to gather and maintain up-to-date information about B's traffic.

Bloom Filters

In order to reduce the amount of exchanged data, a *Bloom filter* [25] is used to represent the contents of a broker's cache (more precisely, the identifiers of cached notifications). A Bloom filter is a space-efficient probabilistic data structure that supports time-efficient membership queries for stored items. Queries might return false positives, but the probability for them can be reduced by increasing the size of the Bloom filter in combination with the number of hash functions used. Thus, there is an inherent trade-off between space and accuracy. Using Bloom filters, it is not possible to deterministically specify the identities of notifications shared by different brokers. Only their number may be estimated which is sufficient here.

Bloom filters are often used in network applications because they can significantly reduce the amount of data sent over the network [113, 150, 161]. A Bloom filter is represented by an array A of bits. The length n of A is called the *size* of the Bloom filter. Initially, all bits of A are set to zero. Besides the bit array, a Bloom filter uses k independent random hash functions $h_1, h_2, \ldots, h_k$ which map each element to be stored or searched in the Bloom filter to a random number uniformly distributed over the range $0, \ldots, n$. An element e is added to a Bloom filter by hashing it with each hash function h_i and setting the bit on position $h_i(e)$ of A to 1. The check whether an element is contained in a Bloom filter works similarly: the element is hashed with each hash function and if any of the bits $A[h_i(e)]$ equals 0, the resulting number is not contained in the Bloom filter, otherwise it is; Algorithm 16 details the procedures for inserting an element into a Bloom filter and testing if an element is contained in a Bloom filter which is also illustrated in Figure 5.3.

Algorithm 16 Bloom filter operations

Contains a bit array A of length n, set of Elements $\mathcal{E}$, and k independent hash functions $h_1, \ldots, h_k$ with

$$h_i : \mathcal{E} \rightarrow \{0, \ldots, n - 1\} \quad \forall i = 1, \ldots, k.$$

1   **procedure** addElement(Element e)
2   **for** i in $\{0, \ldots, k\}$ **do**
3     $A[h_i(e)] \leftarrow 1$
4   **endfor**

5   **procedure** checkElement(Element e)
6   **for** i in $\{0, \ldots, k\}$ **do**
7   **if** $A[h_i(e)] \neq 1$ **then**
8   **return** false
9   **endif**
10   **endfor**
11   **return** true

Due to collisions it is possible that Hash functions set bits of A to 1 for different input values. Thus, false positive results from the containment test are possible,

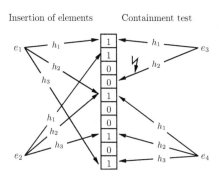

Figure 5.3: The example shows the insertion of elements e_1 and e_2 into a Bloom filter of size 10 using 3 hash functions. Afterwards it is tested whether e_3 and e_4 are contained in the Bloom filter (this test results in a false positive for e_4)

for example, if all the bits that are set to 1 for a particular element e have been set when inserting other elements into the Bloom filter without e having been inserted. Moreover, if more elements are stored in a Bloom filter the number of bits set to 1 increases, raising the probability of false positives thereby. It is obvious that the choice of n and k affects both, processing complexity as well as the probability of false positives in containment tests. Given the number m of elements to be stored in a Bloom filter and the size n of the Bloom filter, Mitzenmacher and Upfal provided a way to calculate the best choice for the number of hash functions k in order to minimize the probability of false positives [114].

Using Bloom filters in the scenario assumed here might appear as an implementation detail which is not important for the overall concept. However, broadcast messages are the basic mechanism the heuristic builds upon and it is, thus, sensible to keep the influence introduced by the heuristic on the system as small as possible. To reach this goal, it is important to minimize the message size of broadcast messages and Bloom filters are an excellent choice in this case. Besides this application, Bloom filters have also been proposed for building hierarchical subscription summaries in publish/subscribe research [150, 161].

Phases of the Algorithm

Brokers regularly broadcast their processing costs and a Bloom filter containing their cache as part of m^K in their neighborhood.

Evaluation Phase. When a broker B_i receives the Bloom filter of another broker B_j, it starts the *evaluation phase*. In this phase, B_i tries to figure out whether it is beneficial from its perspective to connect directly to B_j. This is the case, if B_i and B_j have a significant common traffic, such that the total cost of forwarding and processing notifications is decreased after reconfiguration.

Consensus Phase. If B_i comes to the conclusion that it is sensible to connect directly to B_j, it has to coordinate its request for reconfiguration with the other brokers affected by this reconfiguration. This so-called *consensus phase* is important due to the limited knowledge of the individual brokers. It might happen, for example, that B_i decides for a reconfiguration that is beneficial for itself but raises unacceptable costs for other brokers. In this phase, B_i asks the directly affected brokers about their estimation of the upcoming costs after the reconfiguration and about which link is to be removed in favor of the new link between B_i and B_j in order to keep the topology acyclic.

Reconfiguration Phase. If the reconfiguration still seems sensible after the consensus phase and a link to remove has been identified, the *reconfiguration phase* starts. In this phase, the actual reconfiguration is executed by exchanging the two links in the broker topology, while avoiding notification loss and maintaining message ordering. The mechanism for the reconfiguration has already been described in Section 4.4 and will, thus, not be described here.

In the following, the evaluation and consensus phase are explained in detail for the heuristic called *Cost and Interest heuristic* (CI heuristic).

5.5.2 Phase 1: Evaluation

Based on the information gathered about their local environments, brokers evaluate alternative overlay connections to nodes in their neighborhood. The brokers use a heuristic to determine whether it is beneficial to establish a direct link instead of routing notifications indirectly via intermediate brokers. The heuristic builds upon a basic case involving three brokers, which is described in the following and extended subsequently to larger numbers of nodes.

This phase starts when broker B_i receives a broadcast message $m_{B_j}^K$ which stems from another broker B_j. To illustrate how the heuristic evaluates if a direct connection between B_i and B_j is sensible, a simple example consisting of three brokers B_i, B_j, and B_k is used as a starting point.

Basic Case

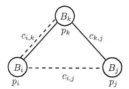

Figure 5.4: Basic case for the heuristic with three brokers

The setup of the basic case is composed of the brokers B_i, B_j, and B_k as shown in Figure 5.4. Let T_1 be the spanning tree (as indicated by the solid lines) which represents the current overlay topology, where B_i and B_j are connected indirectly via B_k. Let T_2 be a possible alternative tree (indicated by the dashed lines) containing a link which directly connects them. Furthermore, it is assumed that B_i received a Bloom filter representing B_j's cache entries, whereon B_i starts the evaluation phase. Let $I(S)$ be the number of identical notifications all brokers in a set S consume (i.e., each broker in S has a local client that either published or subscribed for each notification counted). Using the Bloom filter, B_i can determine

$I(\{B_i, B_j\})$ probabilistically. In the following, B_i's approximation is denoted as $I_{i,j}$ and the same notation is used for analogous estimations.

Given $I_{i,j}$, B_i has to evaluate whether it is sensible to directly connect to B_j. This is the case if the estimated $\text{cost}_{B_i}(T_2)$ of the alternative tree T_2 is lower than $\text{cost}_{B_i}(T_1)$ caused by the current topology. From B_i's point of view, the cost of a tree is calculated as the sum of the communication and processing costs that are caused by routing the notifications both consume (or publish). For the current topology this is

$$
\text{cost}_{B_i}(T_1) = \underbrace{I_{i,k} \cdot (p_i + c_{i,k} + p_k)}_{(a)}
$$
$$
+ \underbrace{I_{i,j} \cdot (p_i + c_{i,k} + p_k + c_{k,j} + p_j)}_{(b)}
$$
$$
- \underbrace{I_{i,k,j} \cdot (p_i + c_{i,k} + p_k)}_{(c)} \tag{5.8}
$$

In Equation 5.8, term (a) describes the costs that are caused by the notifications B_i and B_k both consume. This includes the communication costs of the link they share and the processing costs at both brokers. Term (b) represents the costs of the common traffic of B_i and B_j. As the notifications are routed via B_k, the processing costs at B_k as well as the costs of the links connecting B_i with B_k and B_k with B_j are added. With term (c) the costs for forwarding notifications from B_i to B_j via B_k that all brokers consume (otherwise, they would be counted twice) are subtracted. The cost of the alternative topology is calculated analogous:

$$
\text{cost}_{B_i}(T_2) = I_{i,k} \cdot (p_i + c_{i,k} + p_k) + I_{i,j} \cdot (p_i + c_{i,j} + p_j) - I_{i,k,j} \cdot p_i \tag{5.9}
$$

Since a direct connection is only beneficial if it reduces B_i's costs, the costs caused by the trees of the alternative and current topology (T_2 and T_1, respectively) are compared.

Given that B_i and B_j share common traffic (i.e., $I_{i,j} > 0$), a decision criteria can be deduced that can be used to evaluate the link:

$$\mathrm{cost}_{B_i}(T_2) \; < \; \mathrm{cost}_{B_i}(T_1)$$
$$\Leftrightarrow \qquad c_{i,j} \; < \; \frac{I_{i,j} - I_{i,k,j}}{I_{i,j}} \cdot (c_{i,k} + p_k) + c_{k,j} \qquad\qquad (5.10)$$

The link's communication costs are compared to the costs for routing a notification via an intermediate broker reduced by a fraction proportional to the amount of traffic the intermediate broker also consumes. Hence, the right side of Equation 5.10 can also be interpreted as the communication costs of an indirect connection. If B_i and B_j do not share any traffic (i.e., $I_{i,j} = 0$), the evaluation phase is aborted.

The result is quite intuitive, saying that it is sensible to directly connect B_i with B_j if the costs for forwarding the notifications B_i and B_j both consume is lower when directly connecting B_i with B_j than routing them over B_k. Doing this it is considered that B_k might also consume a subset of these notifications.

But how can B_i calculate $I_{i,j,k}$ if it only knows its own cache contents I_i and the Bloom filters of I_j and I_k? Fortunately, it is possible to create a Bloom filter of the intersection of two sets each represented by a Bloom filter by simply ANDing the bit vectors of both Bloom filters. This way, B_i can compare its local cache against the newly created Bloom filter to calculate $I_{i,j,k}$. The new Bloom filter of the intersection of two sets does not introduce any additional false positives because the resulting bit vector has only bits set that belong to elements in the intersection of both sets. Moreover, the probability of false positives is even reduced this way [111].

Generalization

Up to now, only the basic case has been considered which is limited to one intermediate broker. However, in general, other brokers might be more than two hops away in the overlay topology. To evaluate, whether a direct link to such a broker is beneficial, the right side of Equation 5.10 is used to define the costs $C_{i,j}$ of an indirect connection recursively, based on the path $B_i, \ldots, B_k, B_j$ with $i \neq j$:

$$
C_{i,j} =
\begin{cases}
c_{i,j} & , B_i \in \mathcal{N}_1(B_j) \\
\dfrac{I_{i,j} - I_{i,k,j}}{I_{i,j}} \cdot (C_{i,k} + p_k) + c_{k,j} & , \text{otherwise.}
\end{cases}
\tag{5.11}
$$

Thereby, estimating the costs of an indirect connection of length n is reduced to calculating the costs of a path of length $n - 1$ until the basic case with one intermediate broker is reached. Additionally, a general decision criteria is obtained: a broker B_i prefers a direct link to another broker B_j (that is more than one hop away), if the direct communication costs are less than the calculated indirect costs, i.e., $c_{i,j} < C_{i,j}$.

5.5.3 Phase 2: Consensus

In the previous section, it has been described how a single broker evaluates whether it is sensible to establish a direct link to another broker in its neighborhood. This decision is based on its own local cost-benefit analysis. However, this local decision may lead to an overall increase in costs. To avoid this, the broker seeks a consensus with the other brokers lying on the cycle which would be created by adding the proposed link. This is called the reconfiguration cycle $R = (\mathcal{V}_R, \mathcal{E}_R)$ (cf. Section 4.2.1). Since one edge of the cycle must be removed in order to keep the topology acyclic, all brokers on R are directly affected by a subsequent reconfiguration. Choosing only this subset of brokers limits the overhead for finding a consensus.

After broker B decided to propose to directly connect to another broker with the new link e_n, it asks every broker on the reconfiguration cycle to estimate the costs of the topology that would result from removing a single edge from R. The costs of the reconfiguration cycle R from the perspective of one broker B_i when removing edge e is defined as follows.

$$\text{cost}_R(e, B_i) = \sum_{B_j \in \mathcal{V}_R} I_{i,j} \cdot C_{i,j}^{\backslash e} \tag{5.12}$$

where the cost $C_{i,j}^{\backslash e}$ is calculated on R without e. Accordingly, the aggregated costs of the topology is given by

$$\text{cost}_R(e) = \sum_{B_i \in \mathcal{V}_R} \text{cost}_R(e, B_i) \tag{5.13}$$

Having calculated $\text{cost}_R(e)$ for all edges on R, the edge e_r that shall be removed is determined such that

$$\text{cost}_R(e_r) = \min_{e \in \mathcal{E}_R}\{\text{cost}_R(e)\} \tag{5.14}$$

Since e_r is chosen as the edge whose removal leads to a topology causing the least costs, the maximum benefit is gained for the affected brokers when replacing e_r by the proposed new link e_n. However, if both links are identical (i.e., $e_n = e_r$), the consensus phase is aborted as it then seems to be unfavorable to add e_n to the topology at all.

5.5.4 Integration

In this section, the integration of the heuristics presented in Section 5.4 ff. into the publish/subscribe broker model is described. This covers the dissemination of the Bloom filters as well as the protocol used in the consensus phase.

Cache

Every broker maintains a cache, where it stores the notifications it consumes, i.e., the notifications that match a subscription of a local client or that have been published by a local client. The integration of the cache into the publish/subscribe broker model introduced in Figure 4.9 on page 132 is depicted in Figure 5.5.

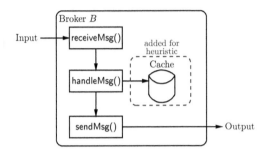

Figure 5.5: Integration of the notification cache needed for the heuristic into the publish/subscribe broker model

To find out about common traffic of a set of brokers it suffices to store only the IDs of the notifications. If notification IDs do not already exist it is fairly simple to implement them. For example, the publishing broker could use a timestamp together with its own unique ID to create a message ID and append it to the notification.

A ring buffer is used to implement the cache. This way, it is guaranteed that the size of the cache is limited. However, it is possible, that the amount of notifications to be stored exceeds the size of the cache. In this case, the oldest messages added to the cache are discarded. Thus, the size of the cache is a compromise between accuracy of the broker's traffic and the storage space needed for the cache.

Broadcast Messages

Every broker B regularly broadcasts the Bloom filter of its cache in the time interval Δt to the brokers in its neighborhood $\mathcal{N}_\eta(B)$. Therefore, B sends a *broadcast message* m_B^K to all its direct neighbors, it shares a link with. The broadcast contains a *time-to-live* (TTL) counter initialized with η and the Bloom filter representing B's cache entries. Additionally, it stores the path including the brokers and links the message has already passed together with the processing and communication costs, respectively.

On receiving a broadcast message, broker B handles it as described in Algorithm 17. First, B determines its common traffic with the sender by comparing its cache contents to the Bloom filter stored in the message (cf. line 6). Then, it updates its knowledge about its neighborhood regarding the communication costs, the processing costs, and the Bloom filter stored in m^K (lines 9–20). In the following, B estimates if it is beneficial to directly connect to the originator of the broadcast message. If this is the case, it sends a request message to the brokers on the path from itself to the originator (lines 23–26). The method used to estimate the benefit of a direct connection depends on the heuristic used. Details for the different algorithms were already described in the previous sections. Finally, B forwards the message to all neighbor brokers that are not already listed in the path stored in the broadcast message given that the TTL is not already zero, i.e., the length of path is smaller than η (lines 29–37). Before doing this, B updates the message by appending itself to the path, the communication costs of the link the broadcast was received over, and its processing costs.

With every broadcast message a broker receives, it learns about its environment (i.e., the processing and communication costs as well as the common traffic). Information about brokers in the neighborhood of a broker B may become obsolete when reconfigurations changed the topology in a way such that these brokers do not belong to the neighborhood anymore. Therefore, stale information is removed by a garbage collection algorithm based on a configurable *update interval*.

In the beginning or due to a reconfiguration, the information about the neighborhood may not be sufficient for B_i to calculate $C_{i,k}$ for every broker B_k on the path to B_j it receives a broadcast message from. In this case, B_i cannot decide whether a reconfiguration is sensible and, thus, has to wait until it has gathered enough information about its neighborhood. Meanwhile, broadcast messages are still forwarded regularly, but subsequent evaluation or consensus phases are aborted.

Algorithm 17 Handle broadcast message m^K

---

1 Set **procedure** handleBroadcastMsg(Broker B_{sender}, BroadcastMsg m^K)
2 **begin**
3      $\mathcal{M} \leftarrow \emptyset$
4   path $\leftarrow m^K_{\text{path}}$
5   // *determine and store common traffic*
6      $I_{B_{\text{this}}, \text{path}[0]} \leftarrow |\{n \in \text{cache} | n \in m^K_{\text{BloomFilter}}\}|$
7      $i \leftarrow 0$
8   // *update knowledge about brokers in the neighborhood*
9   **while** $i < |\text{path}|$ **do**
10         $B \leftarrow \text{path}[i]$
11   env $\leftarrow$ env $\cup\, B$ // *add broker to environment*
12   env$(B)_{\text{BloomFilter}} \leftarrow B_{\text{BloomFilter}}$ // *store Bloom filter of broker*
13   env$(B)_{\text{p}} \leftarrow B.\text{p}$ // *store processing costs of broker*
14   env$(B)_{\text{lastUpdate}} \leftarrow$ now // *store time of last update*
15   // *store communication costs*
16   **if** $i < |\text{path}| - 1$ **then**
17   env$(c_{B,\text{path}[i+1]}) \leftarrow m^K_{c_{B,\text{path}[i+1]}}$
18   **endif**
19         $i \leftarrow i + 1$
20   **endwhile**
21   // *request reconfiguration if beneficial (according to the heuristic used)*
22   **if** directConnectionBeneficial(path) **then**
23         $m^Q \leftarrow$ new RequestMsg()
24         $m^Q_{\text{path}} \leftarrow \langle B_{\text{this}}, \text{path}[n-1], \ldots, \text{path}[0]\rangle$
25         $m^Q_{\text{cost}} \leftarrow \langle 0, \ldots, 0\rangle$
26   handleMsg(m^Q)
27   **endif**
28   // *forward message to neighbors*
29   **if** $|\text{path}| < \eta$ **then**
30   // *append to path and add processing and communication costs to m^K*
31         $m^K_{\text{path}} \leftarrow$ append(path, B_{this})
32         $m^K_{\text{p}} \leftarrow \langle \text{path}[0].\text{p}, \ldots, \text{path}[n-1].\text{p}, B_{\text{this}}.\text{p}\rangle$
33         $m^K_{\text{c}} \leftarrow$ append($m^K_{\text{c}}, c_{\text{path}[|\text{path}|-1], B_{\text{this}}}$)
34   **forall** $B \in \mathcal{N} \setminus \{\text{path}[0], \ldots, \text{path}[n-1]\}$ **do**
35             $\mathcal{M} \leftarrow \mathcal{M} \cup (B, m^K)$
36   **endfor**
37   **endif**
38   **return** $\mathcal{M}$
39 **end**

Request Messages

While broadcast messages serve to gain knowledge about the local environment and are, thus, a prerequisite for the evaluation phase, *request messages* coordinate the consensus phase to finally decide whether the introduction of a new link is beneficial and which link has to be removed in turn. When broker B_i decides to add a new link connecting it with broker B_j, its decision is based on the last broadcast message it has received from B_j. Thus, the reconfiguration cycle R consists of the path the broadcast message was forwarded along and the new link which directly connects B_i and B_j.

Broker B_i starts the consensus phase by sending a request message to B_j along the reverse path of the broadcast message. The request message consists of a cost vector containing an element for every link in R. After receiving the request message, each broker on R adds its own costs for every link on R it has calculated according to the heuristic used (lines 5–7; cf. to Equation 5.12 for the formula of the CI heuristic). Then, the request is forwarded to the next neighbor broker on R until it reaches B_j (lines 9–12). After adding its calculated costs, B_j examines the resulting cost vector to determine the "most expensive" edge, whose removal leads to the best solution for all brokers on R. If this is the proposed new link between B_i and B_j, B_j discards the reconfiguration. Otherwise, B_j starts the reconfiguration phase (cf. Section 4.4) (lines 14–21).

It is possible that the next broker on the path in the request message is not a neighbor of B_j anymore. This can happen because of parallel reconfigurations. In this case, B_j simply drops the request message (line 10).

5.6 Evaluation

A simulation-based approach is taken to compare the CI heuristic with the other two heuristics presented in the previous sections. The advantage of simulation in contrast to formal analysis is that the complexity of the scenarios do not have to be simplified to a great extend. On the downside, running simulations is time-

---

Algorithm 18 Handle request message m^Q

---

1 Set **procedure** handleRequestMsg(Broker B_{sender}, RequestMsg m^Q)
2 **begin**
3     $\mathcal{M} \leftarrow \emptyset$; path $\leftarrow m^Q_{\text{path}}$; $n \leftarrow |\text{path}|$; $i \leftarrow 0$
4     $R \leftarrow < \{B_0, B_1\}, \ldots, \{B_{n-2}, B_{n-1}\}, \{B_{n-1}, B_0\} >$ with $B_i = \text{path}[i]$
5   **forall** $e \in R$ **do** // add cost for removing one edge from R
6        $m^Q_{\text{cost}[i]} \leftarrow m^Q_{\text{cost}[i]} + \text{cost}_R(e, B_{\text{this}})$
7   **endfor**
8   **if** path$[n-1] \neq B_{\text{this}}$ **then** // forward request message along the path
9        $B \leftarrow \text{path}[i]$ with $B_{\text{this}} = \text{path}[i-1 \mod n]$
10   **if** $B \in \mathcal{N}$ **then** // check if topology has not changed meanwhile
11           $\mathcal{M} \leftarrow (B, m^Q)$
12   **endif**
13   **else** // evaluation finished
14        $\min \leftarrow i$ with $m^Q_{\text{cost}[i]}$ is minimal
15   **if** $\min \neq n-1$ **then**
16   // consensus about removing old edge $\{B_i, B_{i+1 \mod n}\}$
17           $m^l \leftarrow$ new LockMsg()
18           $m^l_{\text{path}} \leftarrow \langle B_{n-1}, \ldots, B_0 \rangle$ with $B_i = \text{path}[i]$
19           $m^l_{\text{edge}} \leftarrow \langle B_i, B_{i+1 \mod n} \rangle$
20   handleMsg(B_{this}, m^l)
21   **endif**
22   **endif**
23   **return** $\mathcal{M}$
24 **end**

---

consuming and it is cumbersome to study the effects of parameter variations. For the simulation, a discrete event simulator is used. A new simulator was implemented for various reasons. The most important reason was that most available simulators still need a considerable amount of programming to make it suit to this application's needs [119]. Another reason was that having an own simulator one is aware of everything happening inside the simulator which is important for interpreting simulation results.

The CI heuristic was implemented and decentralized versions of the MST and the MIT heuristic (called ℓMST and ℓMIT) described in Section 5.4 were chosen for comparison. Like the CI heuristic, they both rely on broadcast and request messages. While ℓMST only considers communication costs, ℓMIT concentrates solely on the notification traffic and is, thus, close to the approach by Baldoni et al. [15]. Several experiments are conducted to compare the three heuristics.

The goal of the experiments is to evaluate the following properties of the heuristics:

- the relation between the costs caused by the CI heuristic and the savings gained by reconfigurations (cost-benefit analysis),

- their ability to adapt the system to changes in the environment,

- the performance of the CI heuristic compared to ℓMST and ℓMIT and the lower bound that is theoretically possible,

- the effectiveness with varying heterogeneity regarding how many clients connect to each broker,

- the characteristics of the resulting topologies,

- the number of reconfigurations carried out by the heuristics,

- the effect on the performance of the different heuristics when changing the weights of the communication and processing costs, and

• the influence of locality of subscribers with respect to their publishers.

Some plots of the results show the theoretical lower bound for the costs where sensible. Since it is hard to find an optimal solution for a given distribution of clients and a given broker overlay network, it was decided to calculate the minimum costs that are necessary to distribute the notifications published. The minimum spanning tree in the broker overlay network consisting only of the affected brokers for each notification n published (i.e., $V(n)$) is determined and the costs are summed up accordingly. This lower bound is purely theoretical because it assumes that every notification published is disseminated over its very own broker sub-topology leaving aside the fact that it is assumed here that there is only *one* single broker tree which is responsible of disseminating all notifications. However, this gives a rough estimate about the lower bounds that are possible.

If not stated otherwise, measurements are conducted every 1000 simulation ticks. Each point in a plot is the result of 50 simulation runs to achieve reasonably small confidence intervals. Reconfigurations were carried out without any ordering guarantees. Since ordering guarantees do not induce any additional traffic, this has no effect on the results.

5.6.1 Simulation Settings

This section discusses two sets of parameters. The first set describes the scenario which serves as a testbed for the heuristics. The second set describes the parameters of the heuristics.

Simulation Scenario

The proposed heuristics are evaluated by conducting simulation experiments. For creating physical network topologies that are needed to gain the communication costs the transit-stub model is used which produces Internet-like topologies. In particular, BRITE [109] is used to create 50 different realistic Internet-like topologies for the experiments, each with 100 domains and over 10,000 nodes (the parameterization of BRITE is described in Appendix A). In all simulations 100 brokers are placed randomly on each network topology. This number is restricted by the capacity of the computers that ran the simulation. The publish/subscribe behavior is modeled by introducing 50 different types of notifications or *jobs*, each produced by one publisher and subscribed for by subscribers connected to 9 different brokers. It is important to note, that it is prevented that two subscribers of one notification type are assigned to the same node. From the system perspective it does not matter if there is one or more subscribers connected to one broker, since a notification is routed there only once in either case. This approach has been taken to create reasonably complex message flows by superposition and model a realistic usage scenario this way.

The distribution of clients to brokers is chosen probabilistically according to a load-value that is either fixed ("uniformly distributed clients" in the first experiment) or randomly chosen. The rationale behind assigning different load values is that it is quite common that some brokers attract more clients than others. Introducing this notion of structure or pattern to the system using different load values also gives an impression which algorithms can exploit or adapt to it in order to improve the system performance. A broker is chosen as the local broker for a client with a probability of its load divided by the sum of the loads of all brokers in the system.

Publishers produce on average one notification in five simulation ticks and the publications are exponentially distributed. Broadcast messages are sent every 250 simulation ticks. Thus, every job produces on average 50 notifications per broadcast period.

Table 5.1 gives a concise overview of the scenario parameters used and their default values. In the following experiments, these default values will be used if not mentioned otherwise.

215

Parameter Name	Description	Default Value
numberOfBrokers	Number of brokers in the network topology	100
numberOfJobs	Number of jobs distributed to the brokers	50
numberOfSubscribers	Number of subscribers per job (does not include the publisher)	9
publicationInterval	Average time interval between two consecutive publications	5
publicationDistribution	Distribution of the publications	Exponential
reassignmentInterval	Average time interval between reassigning a job	2500
reassignmentDistribution	Distribution of the reassignment of jobs	Exponential
minProcessingCosts (maxProcessingCosts)	Minimum (maximum) processing costs of one message	0 (10)
minCommunicationCosts (maxCommunicationCosts)	Minimum (maximum) communication costs per link for one message	0 (10)
minLoad, maxLoad	Minimum (maximum) load of a broker, determining its attractiveness to clients	$\langle 1,1 \rangle$ $(\langle 0.1, 1 \rangle)$ for a (non-)uniform placement of clients

Table 5.1: Overview of simulation scenario parameters

Algorithm Parameters

For the experiments, it is necessary to parameterize the algorithms. The configuration of the algorithm is determined by the following configuration parameters:

- broadcastInterval (the time period between issuing two consecutive broadcast messages by one broker)

- updateInterval (the time period after which information of a broker about brokers in the neighborhood becomes stale and is removed)

- environmentSize (the TTL of broadcast messages, i.e., the maximum number of hops a broadcast message is flooded in the broker overlay network)

- cacheSize (maximum number of notification IDs consumed or produced by local clients a broker stores to determine common traffic)

- filterSize (size in bits of the Bloom filter which is used to disseminate the cache contents of a broker in a broadcast message)

- numberOfHashs (number of hash functions used for the Bloom filter)

Finding appropriate values for those parameters is often a trade-off. A low value for broadcastInterval, for example, allows smaller caches and Bloom filters and enables faster reaction to changes on the cost of an increased message complexity. The value of updateInterval has to be chosen with respect to the value of broadcastInterval. It is sensible to set it to a value of broadcastInterval plus the maximum communication and processing delays of a message with respect to the value of environmentSize because it is assumed that no message will get lost. Thus, it is assumed that a broker will receive a broadcast message from each broker in its neighborhood in a time period of length broadcastInterval plus an upper limit of the expected time needed for processing and forwarding a message. The values broadcastInterval = 250 and updateInterval = broadcastInterval + environmentSize · 10 + 1 are chosen. Both values fit well for this setting.

Determining a good value for environmentSize is more difficult because it depends on the scenario and changing it has a significant impact on the overhead induced by the heuristic. Determining a good value for this parameter is, thus, the goal of a separate experiment (Experiment 1).

The same reasoning applies to determining a good value for cacheSize and the respective parameters related to the Bloom filter. Details are discussed in Experiment 2 in the next section.

5.6.2 Determining Heuristic Parameters

Experiment 1: TTL of Broadcast Messages

The TTL of the broadcast messages sent out by every broker is a crucial value for the performance of the heuristics. Increasing this value gives every broker knowledge about a bigger part of the whole network on the cost of an increased message complexity because broadcast messages are forwarded to every neighbor except for the sender. Its value will be chosen according to this experiment, where the reconfiguration costs and the forwarding costs for different values of environmentSize are evaluated to find out about the influence of this parameter on the results.

The setting described in Table 5.1 is used but clients are assigned only once in the beginning, whereby a non-uniform placement of clients is used. At the beginning of each simulation run, a random acyclic broker topology is created. In the first 1000 simulation ticks, the broker topology is left unchanged and the costs of message forwarding is measured. Then, the CI heuristic is started and, after a time period of 29,000 ticks, the costs which were produced in the next 1000 ticks are measured. The costs of regular message forwarding, the costs for forwarding broadcast messages, and the costs induced by reconfigurations are measured separately. The clients are assigned to the network only once in the beginning in order to prevent additional costs introduced by the reassignment of jobs to clients which would influence the costs measured in a misleading way. Each simulation run is repeated 50 times for one setting of environmentSize (ranging from 3 to 12). The value cacheSize = 8192 is used (and filterSize = 98, 304 to achieve a low false positive probability of approximately 0.0033). As one will see in the next experiment, this value is much higher than actually needed. By choosing such a big cache size it is possible avoid side-effects due to false-positives in Bloom filters. The update interval is set to updateInterval = broadcastInterval + environmentSize · 10 + 1, to enable brokers to timely remove information about brokers that vanished from their neighborhood due to reconfigurations.

Figure 5.6 shows that the CI heuristic performs best for a TTL of 3. For values bigger than that, the resulting costs grow and the performance increase drops. This is due to the increased cost caused by the heuristic—mainly due to broadcast messages that are forwarded further in the broker topology. The costs caused by reconfigurations remain very low and are negligible.

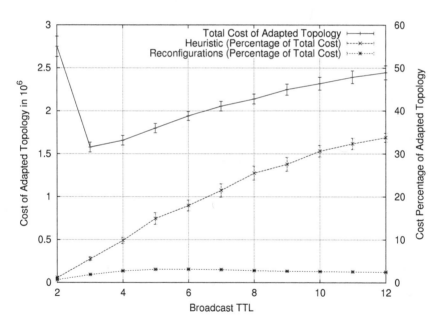

Figure 5.6: Distribution of costs for the CI heuristic with different values of environmentSize

The value environmentSize $= 3$ is set in the following experiments because the results proved best. Accordingly, updateInterval $= 281$ is set.

Experiment 2: Cache Size

Every broker maintains a cache which it regularly distributes in a broadcast message using a Bloom filter. Choosing a sensible size of the cache depends on the publication rate and the distribution of the clients. If it is chosen too small, then the estimation of the heuristic becomes inaccurate. If it is chosen too big, then bandwidth is wasted. In this experiment, the cache size and the Bloom filter size is varied, accordingly. To obtain a low false positive probability of approximately 0.0033, 7 hash functions are used and the size of the Bloom filter is set to twelve times the size of the cache (filterSize $= 12 \cdot$ cacheSize).

The same settings are used as in the previous experiment: the CI heuristic is started and, after a time period of 29, 000 ticks, the costs that were produced in the next 1000 ticks are measured. To test different cache sizes, cacheSize$(x) = x \cdot 256$ is used. By varying x from 1 to 35 and measuring the cost improvement, one can observe how the overall performance increase develops.

To find out about variations in the results of the experiments, the same seeds are used for the random number generators for different cache sizes. To find out about the effect of the cache size on the results, the difference between two consecutive runs with a cache size difference of 256 are then calculated. This is repeated 50 times and the first cache size is used, where the difference in the results is 0 for every run.

Figure 5.7 shows for size x on the x-axis the absolute difference of the results obtained when running the algorithm with a cache size of x and $x + 256$. An optimal value for the scenario here is reached when the curve hits the x-axis for the first time. This is achieved for $x = 18 \cdot 256$, i.e., a cache size of cacheSize $= 4608$ and, thus, filterSize $= 55, 296$.

In the following experiments, the values for cacheSize, filterSize, and

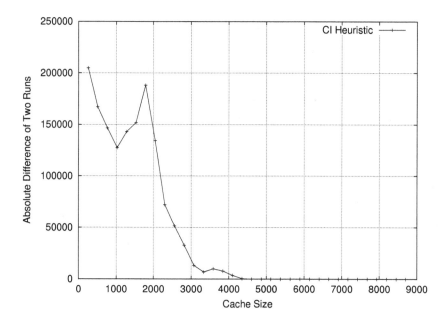

Figure 5.7: Difference in the resulting costs between two consecutive simulation runs with a cache size difference of 256

numberOfHashs determined in these experiments are used. Since those values are smaller than the ones chosen for the previous experiments, it is for sure that this choice does not have an influence on the results gained in the experiments. Table 5.2 gives an overview of the parameters used for the heuristics in the following experiments if not mentioned otherwise.

5.6.3 Performance

The experiments in this section deal with the system costs produced when applying the heuristics. The resulting costs are compared to the costs produced by a random static broker overlay. The experiments comprise uniformly and non-uniformly distributed clients as well as sudden and slow changes of the costs in the network

Parameter Name	Description	Default Value
broadcastInterval	Time between two consecutive broadcast messages sent by one broker	250
updateInterval	Time period after which information about brokers in the neighborhood gets stale and is removed	281
environmentSize	Size of the neighborhood in terms of hops	3
cacheSize	Maximum number of notification IDs that can be stored in the cache	4608
filterSize	Number of bits used for the Bloom Filter	55, 296
numberOfHashs	Number of hash functions used in the Bloom filter	7

Table 5.2: Overview of algorithm parameters for simulation

to find out about the ability of the heuristics to adapt the broker overlay to changes in order to lower the overall costs.

Experiment 3: Performance With Uniformly Distributed Clients

The goal of the heuristics is to lower the total cost of the system. Therefore, all heuristics reconfigure the broker overlay topology based on their own local estimations. In this experiment, the performance of the heuristics in terms of lowering the costs of message forwarding is explored.

50 different topologies are set up, each with 100 brokers randomly placed on nodes in the topology. Clients and processing costs are assigned to brokers with a uniform distribution. For the assignment of the communication costs, the costs provided by the underlying topology are used which are scaled to the same factor as the processing costs (i.e., minProcessingCosts = minCommunicationCosts = 0 and maxProcessingCosts = maxCommunicationCosts = 10). Thus, processing costs have approximately the same impact on the message forwarding costs as communication costs.

Figure 5.8 shows that the costs of the static randomly generated topologies are very high and relatively constant. In contrast to this, all three heuristics are able to lower the costs significantly. The ℓMIT heuristic performs worst because it tries to connect brokers according to the traffic they share. Although this heuristic is able to lower the costs it cannot reach a performance as good as the other heuristics in this case, because it does not exploit the structure of the communication costs which would be beneficial in this scenario with a uniform placement of clients. This is the reason, why the ℓMST heuristic performs better than the ℓMIT heuristic. In contrast to this, the CI heuristic is able to improve the costs even more by considering both, communication and processing costs. It reaches the lowest costs of all three heuristics around $t = 12,000$.

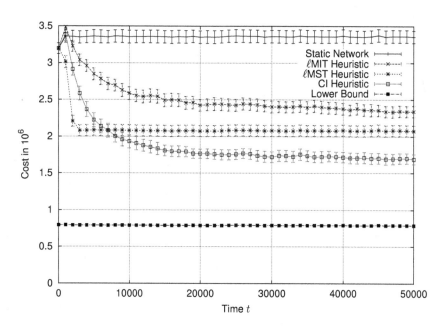

Figure 5.8: Performance of the heuristics in comparison to the static case and the lower bound with uniformly distributed clients

This experiment shows that all heuristics are able to significantly reduce the costs of the system in whole. The CI heuristic performs best although the ℓMST heuristic is able to adapt faster than the other heuristics. The CI heuristic is able to lower the total costs to nearly half of the costs of the random static network.

Experiment 4: Performance With Non-Uniformly Distributed Clients

In this experiment, the distribution of clients is changed compared to the preceding experiment. Therefore, a pattern with respect to the number of clients connected is introduced. The goal is to find out about the ability of the heuristics to exploit this structure to lower the costs of message forwarding in whole.

A *load value* between minLoad = 0.1 and maxLoad = 1 is randomly assigned to each broker. The load value determines the probability that a client is assigned to this broker. This way, "hotspots" are created, i.e., brokers that attract more clients than others. The other settings are equal to the preceding experiment.

The plot of the results of this experiment in Figure 5.9 shows that both the CI and the ℓMIT heuristic are able to exploit the non-uniform distribution of clients to brokers in order to improve system costs. This is due to the fact that some brokers host more clients than others. Those brokers are likely to share a lot of traffic. Thus, it is beneficial to connect those brokers, which both heuristics strive for. The results of the ℓMST heuristic is not significantly affected by the changed client distribution because it does not take common traffic into account and optimizes only based on the communication costs. Accordingly, the results do not differ significantly from the results of the previous experiment.

Again, the CI heuristic performs best in comparison with the other heuristics although the ℓMST heuristic is still able to adapt faster than the other heuristics. The CI heuristic is able to lower the forwarding costs to even more than half of the costs of the random static network. This is possible because the clients are not distributed uniformly and the CI heuristic (as well as the ℓMIT heuristic) is able to exploit this structure in client distribution, where some brokers share more common traffic than others, by directly connecting them.

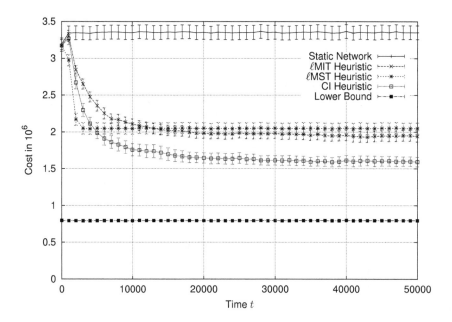

Figure 5.9: Performance of the heuristics in comparison to the static case and the lower bound with non-uniformly distributed clients

It is very likely that the distribution of clients to brokers is not uniform in real-world setting. Thus, the non-uniform client distribution is used in the following experiments according the load value as introduced in this experiment.

Adaptivity

The results of Experiment 3 and 4 show to what extend the heuristics are able to adapt the topology in order to lower the costs of forwarding notifications. As stated earlier in this chapter, network dynamics is a main concern. Thus, it is interesting to evaluate how the heuristics react to changes in the costs and if they are able to cope with the new scenario after they optimized the topology for the previous cost setting. The focus is, hence, on two different scenarios: in the first

(Experiment 5), the costs and client structure are changed suddenly while changes in the second one (Experiment 6) are executed slowly over time.

Experiment 5: Abrupt Changes. To find out about the ability of the heuristics to react to changes in the cost and load values of the brokers, a dedicated physical topology $\mathfrak{T}_0$ with given communication costs is used as at the start of the experiment. At time $t_1 = 25,000$ and $t_2 = 50,000$, the underlying physical topology (i.e., the costs of links) and brokers together with the randomly assigned load values is completely changed. The topology is changed to $\mathfrak{T}_1$ and $\mathfrak{T}_2$ at time t_1 and t_2, respectively, for all simulation runs. Each of the 50 simulation runs per heuristic is initialized with different random seeds such that everything is varied besides the communication costs between the brokers which stem from the underlying physical network topology (i.e., $\mathfrak{T}_0$, $\mathfrak{T}_1$, and $\mathfrak{T}_2$).

Again, all heuristics are able to significantly lower the costs by adapting the broker overlay topology according to the results depicted in Figure 5.10. However, changing the communication costs at t_1 and t_2 results in a spike of the cost-curves of the ℓMST and CI heuristic. The reason for this is that both heuristics take the communication costs into account and optimize the topology accordingly. On the downside, the resulting topology is tailored to the particular communication costs according to the underlying physical network. Changing these costs accordingly, thus, leads to higher forwarding costs.

It is an interesting result that changing the costs and client structure in the network has only a limited impact on the costs that result when applying the ℓMIT heuristic. Obviously, this heuristic does a good job in adapting the topology to lower the costs which is not very sensible to changes in the environment. On the downside, the performance of the ℓMIT heuristic is worse than that of the others or at most as good as the ℓMST heuristic.

Experiment 6: Slow Changes. Since it is not expected to happen that all costs and load values change from one moment to the other, another experiment is

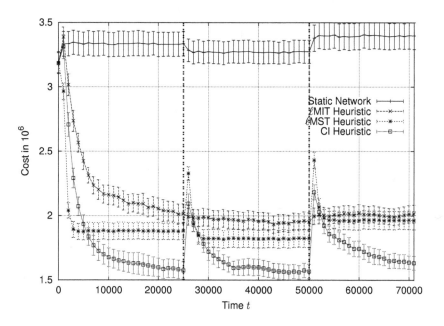

Figure 5.10: Performance of the heuristics in the face of sudden changes in comparison to the static case

performed similar to Experiment 5, where the costs changed linearly from the initial situation to the new one starting at tick $25,000$, stopping at tick $55,000$. Thereby, each cost and load value is changed every 1000 ticks with a constant amount such that the new cost and load values are finally reached at tick $55,000$. In this experiment, $\mathfrak{T}_0$ is also used at the beginning of the experiment and then the topology (i.e., the communication costs) is slowly changed to $\mathfrak{T}_2$ while the other values are chosen randomly as described in Experiment 5.

The results of the experiment depicted in Figure 5.11 show that all heuristic are also able to adapt to slow changes. Interestingly, the slow change for this combination of cost changes is more favorable for the ℓMST heuristic such that it is able to lower the cost to a value below the one it reaches at the end of the

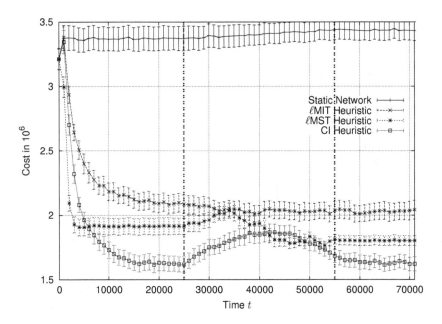

Figure 5.11: Performance of the heuristics in the face of slow changes in comparison to the static case

previous experiment (cf. Figure 5.10). At that point, the limited visibility of the broker overlay that may state a problem for the ℓMST heuristic in the experiment with sudden changes, may be compensated by the slow gradual changes.

Conclusions. All three heuristics are able to cope with dynamic changes in the environment. Their ability to lower the costs depends on the concrete underlying network topology. The topologies created by the ℓMST and the CI heuristic are to a high degree tailored to the underlying network topology because they both take communication costs into account when drawing their reconfiguration decisions. The ℓMIT heuristic performs worst but its cost curve is rather stable against changes in the network.

It is important to note that these experiment have been conducted with only three (two) different physical network topologies. It may, thus, happen that the results are different for other concrete topologies. The goal of this experiment was to explore if the heuristics are able to cope with changes in the network. For comparing the total performance of the heuristics it is more sensible to refer to Experiment 3 and 4, where each simulation run is based on a different physical network topology (50 in total).

5.6.4 Characteristics

The experiments in this section deal with characteristics of the heuristics and the resulting topologies. In the first experiment (Experiment 7), the topology characteristics in terms of broker degrees are measured in order to get an impression of the structure of the broker overlay topologies that are created when applying the heuristics. In the second experiment (Experiment 8), the number of reconfigurations carried out by the heuristic are measured to find out if they are able reach a stable topology.

Experiment 7: Broker Overlay Topology Characteristics

The previous experiments showed that the heuristics were able to significantly improve the costs. It is, however, unclear how the structure of the resulting topologies looks like. Thus, the broker degrees are explored in this experiment which gives an insight in the general structure of the topologies.

For this experiment, the broker degrees are dumped at the end of each simulation run of Experiment 4. Then, the average number of brokers are calculated for each broker degree and heuristic.

The distribution of the broker degrees as plotted in Figure 5.12 shows that all heuristics create a topology, where few brokers have a high degree and the vast majority has a small number of neighbors (i.e., 1 or 2). The topology structure that results from applying the ℓMST heuristic correlates to the structure of the underlying physical network which were generated using BRITE. This network follows a heavy tail distribution as stated in Table A.1 on page 262.

The CI heuristic creates topologies with the highest maximum broker degree (21), while the ℓMIT heuristic results in the lowest maximum broker degree (18). The ℓMST heuristic, however, lies between both heuristics regarding the maximum broker degree (19).

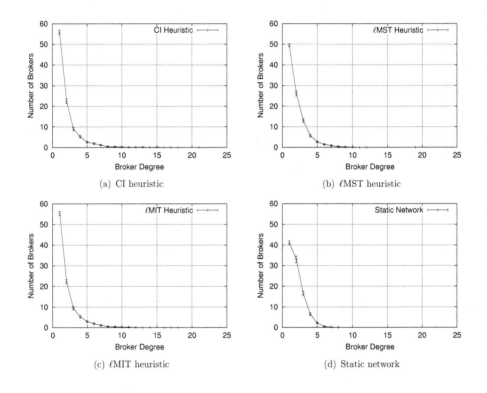

(a) CI heuristic

(b) ℓMST heuristic

(c) ℓMIT heuristic

(d) Static network

Figure 5.12: Broker degree distribution for the heuristics and the static network

The results go in line with the insight that star topologies (or 2-star topologies with at most 2 internal nodes) are in general good solutions for some special cases of the OCST problem like the PROCT problem which are close but not equal to the problem targeted here [157].

Experiment 8: Convergence

All three heuristics try to reconfigure the broker overlay topology in order to lower the costs. Although a dynamic environment is assumed (e.g., clients that are rearranged every reassignmentInterval and publications that are randomly distributed), where parameters change unpredictably, it is interesting, whether the heuristics enter a stable state, where only few to no reconfigurations are executed. In this experiment, thus, the number of reconfigurations carried out by the heuristics are measured, in order to explore their behavior with respect to a stable configuration.

The same settings as in Experiment 5, where the configuration is changed abruptly at $t_1 = 25,000$ and $t_2 = 50,000$, are used. Every 1000 simulation ticks, the number of reconfigurations are measured.

Figure 5.13 shows that the ℓMST heuristic is able to quickly enter a stable state—even after abrupt changes in the costs and client distribution at t_1 and t_2. This is due to the fact that this heuristic focuses solely on the communication costs which are constant in the time periods $[0, t_1]$ and $[t_1, t_2]$. It is, thus, able to quickly enter a stable state.

The CI and the ℓMIT heuristics do not converge to a stable state with respect to the topology and, thus, reconfigurations are permanently carried out. While the CI heuristic shows spikes in the number of reconfigurations at time t_1 and t_2, the number of reconfigurations remain rather constant around 83. The ℓMIT heuristic does not react significantly to the changes in the network but also keeps a rather constant number of reconfigurations around about the half of the configurations carried out by the CI heuristic. This behavior is conform with results gained from measuring reconfigurations in Experiment 5 and 6, where the ℓMIT heuristic proved robust with respect to changes.

It is, however, not a disadvantage in the scenario discussed here to constantly adapt the topology as can be seen in Experiment 5 and 6, where the performance of the CI heuristic is significantly better than that of the ℓMST heuristic and the ℓMIT heuristic.

Figure 5.14 on page 233 shows the reconfigurations carried out by the heuristics

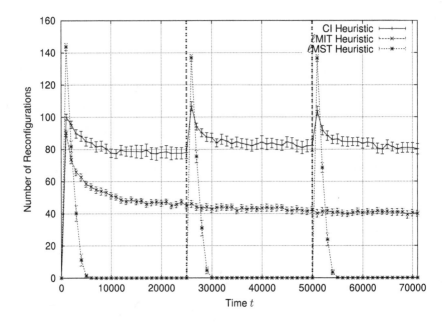

Figure 5.13: Number of reconfigurations in the adaptivity experiment with abrupt changes

for Experiment 6, where the costs are changed in small steps over a time period of 30,000 simulation ticks starting at time 25,000. The cost changes affect the ℓMST heuristic most since the minimum spanning tree changes permanently. The CI heuristic also exhibits a slight increase in the number of reconfigurations. Only the ℓMIT heuristic remains relatively stable. This behavior is similar to the results gained from Experiment 5 with abrupt changes depicted in Figure 5.13.

Regarding stability, the ℓMST heuristic outperforms the other two heuristics. This is no wonder because it relies on stable data which is not changed by the heuristic itself. The CI and ℓMIT heuristic take changing parameters into account like the messages exchanged. Moreover, they work and rely on estimations, where decisions may result in unpredictable consequences. Both are, thus, not able to

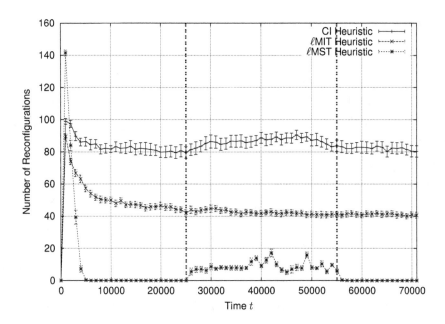

Figure 5.14: Number of reconfigurations in the adaptivity experiment with slow changes

reach a stable state and permanently perform reconfigurations, where the amount of reconfigurations at the CI heuristic is about twice as high as the number of reconfigurations of the ℓMIT heuristic. Although these reconfigurations produce additional costs, both heuristics perform from reasonably well to very well.

5.6.5 Scenario Variations

Most experiments carried out so far were parameterized according to the values described in Table 5.1 on page 216. In this section, it is explored how the heuristics react to different cost weights (Experiment 9) and the influence of locality between publishers and subscribers on the resulting costs are examined (Experiment 10).

Experiment 9: Cost Weights

The previous settings assumed equal weights for communication and processing costs, i.e., minCommunicationCosts = minProcessingCosts = 0 and maxCommunicationCosts = maxProcessingCosts = 10. However, it might be the case that the communication costs dominate the total forwarding costs while the processing costs are negligible or vice versa.

This experiment is parameterized as described in Table 5.1 on page 216 and Table 5.2 on page 222 with minLoad = 0.1 and maxLoad = 1. To model different cost weights, the following values are set maxCommunicationCosts(α) = $(1 - \alpha) \cdot \Gamma$ and maxProcessingCosts(α) = $\alpha \cdot \Gamma$ for $\Gamma = 20$. By varying α from 0 to 1 and measuring the cost improvement, one can observe how the overall costs develop as the processing costs increasingly dominate the forwarding costs and the impact of the communication costs decreases. As in Experiment 1, the cost of message forwarding in a random topology is measured for the first 1000 ticks and the heuristic is started then. At tick 29,000 the costs for message forwarding is measured that accumulate in the following 1000 ticks and the percentage of improvement is calculated.

The results of the experiment plotted in Figure 5.15 shows that the ℓMST heuristic performs worse with a growing α, which reflects the fact that it does only take communication costs into account which are less important with a growing α. The ℓMIT heuristic only concentrates on common interest and performs constantly and even better than the ℓMST heuristic if the processing costs outweigh the communication costs. The CI heuristic performs equally or better than the other heuristics for all values of α as it considers both costs.

This experiment shows that the CI heuristic is superior with respect to different weights regarding the communication and processing costs. It is, thus, more flexible regarding the scenario, where it can be applied. Obviously, the ℓMST heuristics performs well in a scenario, where processing costs are negligible. However, even in this case, the CI heuristic does not perform significantly worse.

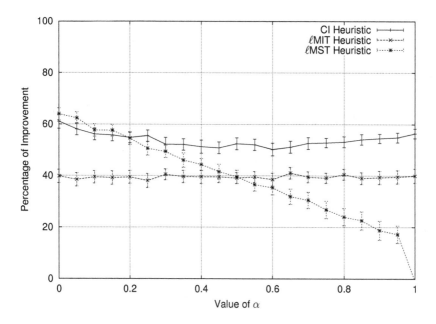

Figure 5.15: Performance of the heuristics for different varying weights of communication and processing costs

Experiment 10: Locality

The experiments regarding the performance of the heuristics for different distributions of clients to brokers already showed that both, the ℓMIT and the CI heuristic, were able to exploit the structure introduced by different client distributions. The non-uniform client distribution thereby modeled the common case that clients are not evenly distributed to brokers. Another factor that has an impact on the distribution of clients that might often be relevant in practice is that of *locality*. In scenarios that exhibit locality, subscribers are located "close" to the publishers which publish the notifications the subscriber subscribed for. The distance metric used in this case often builds on the communication costs between the publisher and the client hosting brokers since this reflects a geographical proximity in many settings.

The setting assumed is similar to the previous experiment with minLoad $= 0.1$ and maxLoad $= 1$ and equal weights of processing and communication costs as described in Table 5.1. In addition to that, a new parameter is introduced which controls the locality called locality. The procedure of assigning the clients of a job to the broker network is changed as follows. The publisher of a job is assigned according to the load values of the brokers. The subscribers, however, are assigned according to a combination of the load value and the locality probability that is determined by the following locality specification, where B_i is the publisher hosting broker and $p_l(i, j)$ is the probability that B_j is chosen as a subscriber hosting broker:

$$p_l(i,j) = \frac{c_{i,j}^{-2}}{\sum_{k=1,k\neq 1}^{\mathcal{N}_B} c_{i,j}^{-2}} \tag{5.15}$$

The probability that a broker is chosen to host a subscriber for a job, where the publisher is connected to B_i, thus, decreases quadratically with increasing communication costs. In this experiment, the parameter locality is used to control the impact of locality on the probability $p_s(i, j)$ that B_j is chosen as the subscriber hosting broker when the respective publisher is connected to B_j in the following way:

$$p_s(i,j) = \text{locality} \cdot \text{load}(B_j) + (1 - \text{locality}) \cdot p_l(i,j) \tag{5.16}$$

By varying locality from 0 to 1, one can explore the whole spectrum ranging from no locality to a subscriber distribution that relies solely on locality.

The percentage of improvement depicted in Figure 5.16 shows that all heuristics exploit the additional structure in the client distribution added due to locality in order to gain only little performance increase. The ℓMST and the ℓMIT heuristic perform about the same and the CI heuristic gains results that are about 10% better. In contrast to the expectations, none of the heuristics was able to gain a significant improvement by exploiting locality.

The results show that the ℓMIT and the CI heuristic are both able to exploit the structure induced by locality. The ℓMST heuristic does also benefit from an

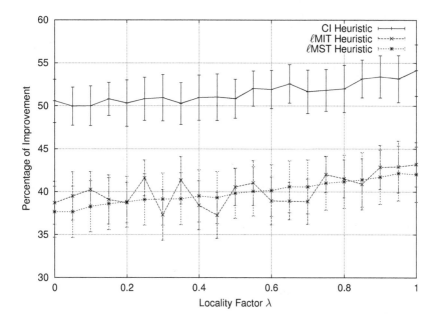

Figure 5.16: Performance of the heuristics for varying locality

increase in locality since the resulting minimum spanning tree topologies connect publisher with subscriber hosting brokers more directly.

5.7 Related Work

Optimizing the publish/subscribe broker overlay topology is important in many scenarios. Moreover, it has an application beyond publish/subscribe systems because the optimization of overlay networks with respect to forwarding and processing costs can also lead to a performance increase for general peer-to-peer systems which often build on overlay networks.

Publish/Subscribe Systems

In recent but preliminary work, Migliavacca and Cugola present an approach that is similar to ours [112]. They base their work on an earlier publication [80] but take a different route to solve the PSOOP than done in [83]. In their approach, every broker optimizes its immediate one-hop neighborhood using an optimization mechanism that builds on a local search technique called *Tabu*. Similar to the approach discussed here, they hope to improve the performance of the system in whole by executing local optimizations. The results are, however, preliminary and solid simulation results are still missing.

Interest Clustering. Recently, there has been an increasing interest in optimizing publish/subscribe systems. Baldoni et al. and Querzoni published more details about their maximum associativity trees approach which was discussed in Section 5.4.2 [17, 132]. The downsides of their approach remain. Moreover, the brokers in their approach exploit the contents of routing tables to exchange common traffic with neighbor brokers which means that their approach is more difficult to integrate into a publish/subscribe system than ours which is independent of the routing algorithm used.

Anceaume et al. present a similar approach which they coin "semantic overlay" [8]. The idea is quite interesting: subscribers cooperatively build dissemination trees according to their subscriptions. As in the previous approach, the basic assumption is that notifications are uniformly distributed in the subscription space and, thus, brokers with large overlapping subscriptions share more common traffic.

Recent work by Baldoni et al. [14] and Chockler et al. [43] considers topic-based publish/subscribe. In contrast to the model used here which builds on one single broker overlay tree for event dissemination, they use one dedicated dissemination tree for each topic. In addition to that, a general overlay network (or membership service) is necessary to route events to their respective dissemination tree. The idea is promising because it is now possible to simply optimize the individual topic trees with respect to their communication costs (i.e., using a minimum spanning tree). However, this approach is only suitable for topic-based routing similar to the approach by Juninger and Lee which relies on a multiple ring topology [89]. Chockler et al. discuss the problem of minimizing the number of pure forwarding brokers for topic-based publish/subscribe systems while maintaining a maximum broker degree [42]. They name this problem *Minimum Topic-Connected Overlay* (Min-TCO) and show that it is NP-hard. Moreover, they present a centralized algorithm that relies on global knowledge and approximates the optimal solution within a logarithmic factor. They prove that no polynomial algorithm can approximate Min-TCO within a constant factor.

Voulgaris et al. present an approach which is very similar to the early work of Baldoni et al. [152, 153]. They cluster brokers based on common interest with respect to similarity in the local subscriptions they manage (a similar approach is taken by Anceaume et al. [9]). The downside of this approach is that it assumes a uniform distribution of events over the event space (or at least over the event space covered by all subscriptions in the system). This assumption does not hold in general as discussed previously.

Quality of Service Management. The work presented here can be seen as part of *Quality of Service* (QoS) management in publish/subscribe systems [20]. In [31], Caporuscio et al. present an approach, where they apply LIRA (Light-weight Infrastructure for Reconfiguring Applications) to the SIENA publish/subscribe system in order to provide QoS guarantees. The approach builds on a separate monitoring component for each broker and client in the system and one centralized application manager component. Threshold values trigger the migration of clients and changes in the broker overlay network topology. The goal is to meet guarantees negotiated previously between participants. Reconfigurations are, thus, triggered when the system may fail to provide certain guarantees. The paper focuses on the architectural implications of integrating QoS management into the SIENA publish/subscribe system. It provides a centralized solution which has obvious shortcomings in large-scale systems.

A distributed approach is taken in INDIQoS, a publish/subscribe System that builds on top of a peer-to-peer routing substrate similar to PASTRY [32]. Its goal is to support QoS, particularly regarding latency and bandwidth, in publish/subscribe communication by enhancing advertisements and subscriptions with adequate extensions. Therefore, brokers in the overlay topology collect QoS information from immediate neighbor brokers and collectively try to provide the requested QoS for subscribers and publishers, respectively. The main benefit is that a significant amount of signaling traffic can be saved in comparison to well-known techniques applied to direct communication, where every node tries to gain up-to-date knowledge of the network or where subscriptions are flooded in order to find an admissible path. Although the system is able to give QoS guarantees this way, there is no mechanism for optimizing the broker topology which is built according to the DHT used.

Peer-to-Peer Routing Substrates

Structured Peer-to-Peer Networks. Structured peer-to-peer routing substrates like CAN [133], PASTRY [136], and CHORD [144] build on *distributed hash tables* (DHT) [91] and provide a simple interface and a scalable way to lookup objects and route to them in a distributed fashion. There have been several efforts to deploy publish/subscribe communication on top of these routing substrates like HERMES [128] and PASTRYSTRINGS [2] based on PASTRY, MEGHDOOT based on CAN [70], and other approaches based on CHORD [148, 149]. Baldoni et al. propose a solution which is even independent of the routing substrate used [18]. Most of these approaches focus on the implementation of powerful matching operations and leave overlay topology issues to the peer-to-peer routing substrate.

In the area of peer-to-peer routing substrates there have been several efforts to adapt the overlay topology to the underlying physical network [39, 44, 92]. However, these approaches mainly concentrate on communication delays which are only one aspect of the underlying network. The authors of [38] even argue that structured peer-to-peer overlays may not be a good choice in dynamic environments. The main problem inherent to all these solutions is that they decouple the optimizations from the actual network traffic in the overlay network that is generated by the application. One approach to combine both is discussed in [110], where the authors try to optimize the overlay topology with respect to communication costs. This approach is a special case of the OCST problem and has been introduced earlier as the MRT problem. The solution proposed is based on local search and builds on a tree structure. Although very interesting, this approach is restricted with respect to the assumptions taken and not suitable to solve the PSOOP in general since processing costs are not considered, for example.

Unstructured Peer-to-Peer Networks. Besides the implementation of publish/subscribe systems on top of structured peer-to-peer networks, there is also work on using unstructured peer-to-peer networks. Unstructured peer-to-peer systems often rely on flooding which can introduce high message complexity. Thus, solutions that build on probabilistic approaches have been proposed [48, 49]. The basic idea is to flood subscriptions only in a limited neighborhood of the subscribers. Notifications are routed according to subscription information if available at the brokers or they are disseminated in a random walk if not. This approach forgoes a single broker overlay topology and the maintenance that comes with it on the cost of only probabilistic delivery guarantees.

Multicast

Publish/Subscribe Over IP Multicast. It is an appealing approach to realize the notification service by a collection of IP multicast trees which, for example, each correspond to a topic. For topic-based routing with a number of topics much larger than the number of available multicast groups, it is in general hard to find an optimal match between the set of brokers and the multicast groups with respect to resource usage (*channelization problem*) [1]. For content-based routing this problem becomes even harder because it is more difficult to cluster brokers into groups. Several approaches have been proposed that try to do an imperfect mapping of interest clusters to multicast groups [156]. In publish/subscribe research, Orpychal et al. published work on the mapping of broker clusters according to their subscriptions to IP multicast groups in a way such that notifications that reach brokers with no subscribed local clients have to drop them [120]. A similar approach is taken by Riabov et al. [134].

Applying IP multicast for publish/subscribe systems works well as long as it is possible to determine distinct groups of brokers that all share a dedicated set of messages which is easy for topic-based routing but in general hard for content-based routing. Accomplishing this perfect mapping can lead to an explosion of the number of groups for content-based routing up to 2^n, where n is the number of clients [120]. Thus, the number of groups has to be bounded such that messages are sent to brokers with no subscribed client as in the approach with only one broker overlay network. In addition to this overhead, extra work must be invested in the maintenance of the multicast groups.

Application-Layer Multicast. Kwon and Fahmy present an algorithm that constructs topology-aware overlay networks for multicast groups [98]. They focus on communication and bandwidth constraints and propose a distributed algorithm that achieves low latency multicast trees. In contrast to the approach presented here, where many implicit multicast groups share one broker overlay topology, their approach considers a unique broker overlay for each multicast group. Besides this, the authors do not consider processing costs which may add additional delay to message forwarding.

Banerjee et al. take another approach to find minimum average-latency degree-bounded spanning trees for building efficient multicast overlay networks [19]. Their architecture called OMNI comprises a set of *multicast service nodes* (MSNs) that are distributed in the network. Clients connect to these MSNs and there is only one publisher connected to an MSN called "root". MSNs form an acyclic overlay network for notification forwarding. To optimize the average latency of notification delivery, the authors exploit the fact that MSNs with more clients have a bigger influence on the average latency. Their distributed approach relies on basic operations in the tree like swapping of MSN positions (parent-child swap) or promoting child MSNs up the tree if an MSN has not yet reached its maximum degree. The distributed approach exploits the hierarchical structure of the dissemination tree. In contrast to the approach discussed in this chapter, the authors focus only on latency and do not take peer-to-peer routing with several publishers into account.

Papaemmanouil et al. extend the approach by Banerjee et al. by adding profile-related as well as cost-related extensibility [121]. The result is a very flexible system which can be configured for various usage scenarios. However, the focus is still on hierarchical routing which is necessary for the algorithm to decide if a certain reconfiguration is beneficial.

The KYRA publish/subscribe systems takes another approach to optimize notification dissemination by using multiple broker trees, each responsible for parts of the notification space and organized related to their network proximity [30]. The partitioning of the notification space is done according to popularity (with respect to subscriptions), the number of notifications published and the resources a notification consumes when being processed. This approach is very promising but not feasible with one single broker overlay topology for notification dissemination.

There are various other approaches that optimize dissemination trees for application-layer multicast, especially with respect to data streaming [166, 167]. They all bear the same basic problem of mapping publish/subscribe communication to multicast systems which have already been discussed in the context of IP multicast above.

Other Approaches to Overlay Optimization

Besides the optimizations discussed so far, there are other approaches that try to optimize a publish/subscribe system with respect to different goals. One approach that builds on top of a model that comprises utility functions for consumers is presented by Lumezanu et al. [104]. The optimizer presented adjusts the publication rates for different consumer classes and determines the number of consumers for each consumer class in order to improve the utility of the system in whole measured with the utility function. The problem is hard to solve, because there are upper limits for different resources (i.e., CPU power and network bandwidth). The optimization is performed according to a "price" which is calculated for each resource and introduces a market-like mechanism.

The availability aspect of overlay optimization has not been discussed so far. A first approach for fault-resilient publish/subscribe broker overlay topologies is presented in [97]. The authors propose an algorithm which creates overlay networks that are able to cope with k independent failures of overlay links between any two brokers. The solution is based on a central and globally known set of *stellar nodes* around which the network is grown. In this network, every broker holds at least k connections to other different brokers. The approach is interesting and introducing fault tolerance considering the physical underlay network is important. Unfortunately, the authors do not discuss the implications on the publish/subscribe layer, for example, on the publish/subscribe routing layer.

5.8 Discussion

In this chapter, the problem of finding an optimal broker overlay topology for publish/subscribe systems in a cost model that comprises communication as well as processing costs was discussed. It was shown that this problem is NP-complete and that using a heuristic to approximate a good topology is a viable option. The shortcomings of two heuristics proposed in the past were discussed and the CI heuristic was presented which takes processing and communication costs into account in order to improve the performance of the system in whole. A framework was proposed, where brokers exchange information in a bounded neighborhood to create a limited local view of the network. This view is then used in order to optimize the topology of the neighborhood of each broker. The framework is independent of the routing algorithm used and is, thus, easy to integrate into any publish/subscribe system that builds on an acyclic broker overlay topology.

In the evaluation, several experiments showed that the CI heuristic is superior in most of the experiments to the other two heuristics ℓMST and ℓMIT. The results show, that all heuristics are able to improve system performance with respect to a random topology which would be created when a publish/subscribe system grows naturally by adding and removing brokers as leafs. The simple localized approach presented emerges in an improved broker topology with lower costs according to the cost model used here. The additional overhead is reasonably small and negligible with respect to the performance gain.

The algorithm framework used for the three heuristics builds on messages broadcasted which are sent out regularly. This produces a constant basic load on the network even if no client initiated messages are forwarded. An alternative approach would be to send broadcast messages after a threshold has been reached in the number of notifications a broker consumed as proposed by the authors of [17]. This, however, poses the question of the minimum and maximum broadcast interval and degrades the reactivity of the system in case of a low publication rate. Moreover, it does not allow the system to quickly react to changes in the communication costs since the optimization algorithm is triggered by the arrival of a broadcast message. Thus, further work would be necessary here. It is in general dependent on the requirements of the application which approach is more suitable.

Some issues remain open for future work. Although the CI heuristic performs very well, it does not reach a stable state with respect to the reconfigurations. Experiment 8 shows that neither the ℓMIT heuristic nor the CI heuristic were able to reach a stable topology, where no reconfigurations are executed anymore—in contrast to the ℓMST heuristic which quickly reaches a stable state. The reasons for that have already been discussed and go back to the estimations carried out by the brokers and the complex dependencies between reconfigurations and the notification flows. Thus, further work is needed to improve the CI heuristic to be able to reach a stable topology.

A general issue is that all heuristics were evaluated and designed with respect to the *average* message forwarding costs. This may not fit to every application domain, where it is conceivable that the topology may need to be optimized with respect to the maximum costs for forwarding one message (e.g., if the costs relate to some fault tolerance metric). It was also assumed that all messages cause about the same cost when being sent or processed. This may not hold in general. These issues are left for future work.

The parameters of the algorithm framework were determined in separate experiments. They depend on the scenario in different degrees which means that the parameters used for the experiments are not optimal in general. Generally speaking, this affects all parameters described in Table 5.2 on page 222, where filterSize and numberOfHashs can be determined using cacheSize as already discussed in Experiment 2. The situation is similar for updateInterval which depends on broadcastInterval. Thus, it remains to determine broadcastInterval as well as environmentSize and cacheSize. For example, it could be sensible to increase the broadcast interval in case of low notification traffic. Similarly, it may be beneficial to increase the cache size in face of a growing number of notifications consumed. In both cases it is necessary that brokers in the neighborhood adapt accordingly. For example, it would be necessary to increase updateInterval for the values of a broker which increased its broadcastInterval. Similarly, brokers must be able to handle different sizes of Bloom filters when calculating the common traffic. Finding these parameters automatically or even adapting them at runtime is an interesting issue for future work.

Finally, it was assumed that changes of the overlay network topology do not affect the underlying topology such as processing capacity of the nodes running the broker and communication links. It is currently an interesting general issue how optimizations on the overlay network can be carried out in a way such that they are not counterproductive to optimizations carried out in the underlying physical network topology [5, 140].

6 Conclusions and Future Work

6.1 Summary

The objective of this book is to approach the goal of rendering publish/subscribe systems self-managing. This issue is of particular interest since most of today's publish/subscribe systems are manually managed which more and more becomes the limiting factor regarding the growth of these systems. Besides this problem of scale, often termed as the *complexity crisis* [93] and frequently mentioned as the driving force behind research on self-X systems, new applications also demand for self-management features. These systems often reside on the other end of the scale and comprise (small) distributed systems that run in an environment, where administrative support is generally not available. Self-management is inevitable in those scenarios which comprise the consumer market (e.g., the e-home) as well as systems that have to operate autonomously (e.g., for monitoring remote areas).

The analysis of related work showed that research on self-management of publish/subscribe systems has just started and that there are still many interesting open problems and opportunities. The focus in the preceding chapters was on general management issues like reconfiguring the notification service at runtime without service interruption as well as automated fault handling with recovery guarantees and adapting the notification service with respect to the usage patterns of the clients. The desire for self-management features in publish/subscribe systems becomes even more important because the event-based computing paradigm including publish/subscribe research gains growing interest in academia as well as industry. This is demonstrated, for example, by a growing range of professional software products and the first conference dedicated to this area [79].

In this work, various contributions to support self-management in publish/subscribe systems were presented. These contributions range from basic mechanisms like seamless reconfiguration of broker overlay networks, which are often used for a distributed implementation of the notification service, to fault management in form of self-stabilization. Ways to render the notification service self-optimizing were presented and tools to formally analyze the performance of a given publish/subscribe system.

Self-Stabilizing Publish/Subscribe Systems

Chapter 3 presented algorithms for rendering content-based routing in publish/subscribe systems self-stabilizing. The algorithms are based on the concept of leasing, where the soft state held in the routing tables of the brokers is regularly refreshed. This way, it is possible to remove stale routing table entries and correct arbitrary perturbations of the routing tables' contents. In Chapter 4, the self-stabilizing content-based routing layer was complemented with a self-stabilizing broker overlay network. It is also possible to combine it with other algorithms which would also render the broker overlay network self-stabilizing. However, the overlay maintenance algorithm has the advantage that it supports reconfiguration and maintains arbitrary correct broker overlay networks.

Besides algorithms that are dedicated to specific routing algorithms (and are, thus, able to exploit their characteristics for an improved performance), a generalized algorithm was presented which supports arbitrary routing algorithms which comply to the definition of a correct routing algorithm used here. Furthermore, the question was discussed whether the overhead induced by self-stabilization is worthwhile and those scenarios were identified in which filtering outperforms flooding. In general, content-based routing does not perform better than flooding, for example, in scenarios where every broker has a local client that is interested in all notifications published. However, notification filtering often does not pay off in those scenario anyway.

One main benefit of self-stabilization lies in the guarantee that the system recovers from arbitrary transient perturbations of soft state as well as links or brokers that crash and come up again. Some self-stabilizing systems are even able to cope with permanent faults if they are able to find back into a correct state like the self-stabilizing broker overlay network presented in this book. The convergence property of self-stabilizing systems implies that they are guaranteed to find back into a legal state if possible when started in an arbitrary state. This feature is of particular interest in the aforementioned application scenarios, where manual administration is not feasible or the system should be put into operation by non-experts.

Seamless Broker Overlay Topology Reconfiguration

It is expected that operating a publish/subscribe system in a dynamic environment requires reconfiguring the notification service for maintenance, growth, or performance optimizations, for example. Although the publish/subscribe communication paradigm is well suited to dynamic environments the issue of reconfiguration at runtime has mostly been tackled from the perspective of faults in the past. Here, changes are unpredictable making it hard to prevent message loss. The existing algorithms were compared with a new algorithm in Chapter 4 that not only prevents message loss in fault-free scenarios but is also able to maintain message orderings which no other algorithm supports yet. It supports various routing algorithms but may need to be adapted for others. This contribution is significant since it allows reconfigurations to be implemented in a transparent manner for the clients.

Moreover, a self-stabilizing publish/subscribe was presented stack in Chapter 4 which comprises a self-stabilizing broker overlay network as well as self-stabilizing content-based routing and is able to seamlessly implement reconfigurations at runtime. The basic problems inherent to reconfiguring self-stabilizing systems were identified. As a consequence, it was not possible to follow the same approach presented for regular publish/subscribe systems. Thus, a central mechanism which relies on a coloring scheme was introduced that coordinates actions on the different layers. This way, it was possible to prevent message loss and provide ordering guarantees during reconfiguration without losing the self-stabilizing property of the system.

Self-Optimizing Broker Overlay Topology

The topology of the broker overlay network which provides the notification service has a significant influence on the performance of the publish/subscribe system. In the past, most research has concentrated on creating minimum spanning trees and using them for the broker overlay topology. Recently, it has been identified that it could be favorable for the performance (in terms of latency) to minimize the number of overlay hops when distributing a notification. A cost model was presented in Chapter 5 which incorporates communication as well as processing costs. Then, the problem of finding an optimal broker overlay topology was formalized and it was shown that it is NP-complete for the static case, where global knowledge is available. In a dynamic scenario, where global knowledge is not feasible, the problem becomes even more difficult. Thus, a heuristic was used to approximate a good solution and the Cost and Interest heuristic was presented, where the brokers use local knowledge gained from their neighborhood to optimize their neighborhood's topology. The rationale behind this approach is that these local optimizations will hopefully eventually emerge in a good global broker topology.

A comprehensive simulation study showed that the CI heuristic is flexible with respect to the costs distribution and dynamic changes, and that is able to significantly reduce the total cost compared to random static broker topologies. It was compared to the classical minimum spanning tree approach and a heuristic which tries to minimize the number of overlay hops. Although both heuristics were able to improve the performance of the system, the CI heuristic almost always outperformed them. Moreover, the experiments showed that they were not as flexible as the CI heuristic because they did not consider communication or processing costs. For the implementation of the reconfigurations proposed by the heuristic, the reconfiguration algorithm presented in Chapter 4 was used.

Formal Analysis

Models and formal analysis are important to create a deeper understanding of the inner workings of systems. This is especially true for research on publish/subscribe systems, where the vast majority of performance evaluations is based on experimental evaluations. In the context of the evaluation of self-stabilizing publish/subscribe systems in Chapter 3, a stochastic analysis based on Markov chains was presented which is applicable in a wide range of scenarios. For scenarios which build on complete trees and hierarchical routing, where subscribers are uniformly distributed to the leaf brokers and identity-based routing is used, closed formulas were provided. For more complex scenarios with arbitrary tree structures and clients connecting to any broker in the system with configurable probabilities, a formalism based on recursive formulas was presented. The formalism covers a wide range of scenarios and allows to analytically determine the performance of publish/subscribe systems, for example, in face of locality.

6.2 Conclusions

We live in a world, where the pace of change increases permanently. Modern computer systems are on the one hand a driving force behind this development but, on the other hand, also have to keep pace with it. Dealing with dynamics requires rethinking the architecture of today's computer systems over and over again. When taking a closer look at our day-to-day life we realize that our actions are often triggered by *events* rather than a strict schedule planned in advance— in contrast to conventional computer systems, where a procedure call results in planned actions which finally lead to the effect that was intended by the caller [76]. It is, thus, no wonder that there is a movement towards *event-based programming*, where the *event* serves as the central abstraction around which systems are built to better support parallel processing in distributed systems as well as multiprocessor and multi-core processor systems [53]. Realizing event-driven systems on the basis of a publish/subscribe middleware is a sensible decision. In the following, lessons learned from this book are summarized.

Reconfigurability

Research on publish/subscribe systems in the past has often neglected dynamic environments with respect to reconfiguration. In Chapter 4, it was argued in favor of integrating seamless reconfiguration of the broker overlay topology into the publish/subscribe middleware. This, however, is only one aspect of reconfigurability. In general, reconfiguration is vital for running publish/subscribe systems in dynamic environments because it allows for adapting the system to changes in the environment. Adaptation is part of (self-)management and crucial in order to gain an optimal performance. Since reconfiguration should not have a negative impact on the service provided by the middleware, it is important that it happens transparently for the clients. This task is far from being trivial as seen in the case of topological reconfigurations. Nevertheless, it is worthwhile since it broadens the class of scenarios, where publish/subscribe systems can be applied.

Usage-Driven Self-Optimization

The middleware layer sits between the operating system and the application and provides transparency to the latter in form of abstract services. Since the middleware is "close" to the application, it is a reasonable place to invest in optimizations that are not application-specific but driven by application demands. In the case of distributed publish/subscribe systems, where an own network routing infrastructure is maintained in form of a broker overlay network and the interface operations are relatively simple, there are lots of possibilities for performance optimizations according to the application behavior. The decision of when and how to optimize the system can thereby rely on many different approaches. One could be, for example, to forecast behavior if it correlates with time, or to learn from the past using techniques from machine learning. In the book at hand, it was assumed that the near future will be similar to the near past, which is certainly true in many scenarios. One important insight is, that event-based systems, which are loosely coupled in general, need a flexible infrastructure because client behavior may change significantly and unpredictably over time. In an enterprise application integration scenario, for example, one event (like a purchase order) may result in a lot of other events (like billing, issuing related orders, etc.) which can be created in kind of a domino effect. In order to cope with sudden changes in the communication behavior, it is important to quickly react to the behavior at the interfaces which can easily be monitored.

Faults

Besides performance optimization, fault handling is another vital issue for the deployment of publish/subscribe systems in practice. In the past, several researchers worked on fault masking which is certainly one important facet of fault handling. With the work presented here, the fault tolerance mechanisms presented in related work were enriched with self-stabilization. Besides the guaranteed stabilization in the face of transient faults, self-stabilization also provides automated initialization. The latter is a management task which may not only be needed when starting the system for the first time, but also in the face of power outages or maintenance. Self-stabilization is, thus, a feature of great value not only with respect to fault handling.

As already recognized in Chapter 4, self-stabilization might come with restrictions which have to be tackled in the context of reconfiguration. However, although it is not easy to integrate reconfiguration into a self-stabilizing system, it is useful in many scenarios and a valuable supplement.

Modeling and Analysis

Most publish/subscribe systems have been evaluated in the past with simulation-based studies. Thereby, a large body of different simulation tools has been used like NS2, JSIM, PEERSIM, CHORDSIM, not to forget the simulation platforms developed from scratch by individual researchers. Besides this variety of simulation platforms, the datasets used for the simulations are often described insufficiently due to spatial restrictions. Both issues make it hard to compare results gained from different systems. At this point, formal analysis can help to generate a deeper understanding because implementation details can be neglected or must be included into the formalism. Besides that, a formalism provides a dedicated space for modeling systems making it easier to understand the differences between various approaches. On the other hand, formal analysis often requires simplifications in the model to be able to formally describe a system. This can be challenging in complex scenarios like the one described in Chapter 5. Formal analysis is an important tool, especially in the area of publish/subscribe systems, where it has yet only been used rarely. However, it cannot substitute simulations in every case.

6.3 Outlook

In this book, approaches were presented which can be employed to render publish/subscribe systems self-managing with respect to certain aspects. The topics overlay reconfiguration, self-stabilization, and overlay optimization, however, only represent selected facets in the area of self-managing publish/subscribe systems. The findings leave room for further research in this field. In the following, research questions which result from the work in this book are discussed.

Analysis and Modeling

The formal analysis introduced in Chapter 3 lays the foundation for further work. The analysis is restricted to identity-based hierarchical routing. Dropping these restrictions would significantly increase the scope of systems that can be analyzed with the formalism. Moreover, it could be sensible to add processing and communication costs to the analysis instead of solely counting the number of messages produced. This extension is rather straightforward and easy to integrate, similar to the integration of advertisements and peer-to-peer routing.

Extending the analysis goes hand in hand with the question of how to model publish/subscribe systems. This regards aspects like locality, faults, and costs. Having a general model makes it possible to *instantiate* a concrete system that consists of a broker topology and a set of distributions to be analyzed. Thus, it is easier to study the effects of different scenarios and design decisions.

Modeling and analysis opens doors for performance prediction and adaptation. Data collected by monitoring a publish/subscribe system could be used to derive a model which can be analyzed subsequently. Based on the analysis and the model it is possible to derive performance predictions and adapt the system accordingly.

Simulation and Comparability

Besides the analysis, simulation is still an important tool to evaluate and compare implementations of different concepts. However, the research community is still missing widely accepted standard benchmarks concerning cost models as well as subscription and publication distributions. Different application scenarios have different requirements and it is of significant importance to formulate them in order to be able to evaluate and compare different systems similar to benchmarks used in industry, like SPECjms [137]. Identifying scenarios and modeling them accordingly would be of great benefit for the publish/subscribe research community.

Low Layer Publish/Subscribe

In this book, the publish/subscribe communication paradigm was considered from the middleware perspective, where the broker overlay network is layered on top of a physical network topology. Taking a step back and looking at the big picture it is only consequent to push content-based addressing as in content-based routing further down the network stack in order to realize what is sometimes called a *content-based network* [35]. Companies like SOLACE SYSTEMS[1] already provide content-aware routing in hardware—however, this topic has essentially been neglected in academic publish/subscribe research yet. Using content-based networking as a basic networking paradigm could ease the development, deployment, and integration of distributed applications which often enough rely on dedicated overlay networks.

Broker Overlay Management

All work in this book relied on an architecture, where the notification service consists of one dedicated acyclic broker overlay network. With a large set of diverse applications it might not be sensible to use only one topology for notification dissemination because a large set of superposed message streams may leave only little room for optimizations. Various approaches previously discussed as related work present solutions that rely on multiple trees for the dissemination of notifications. This approach raises new problems in dealing with duplicates and ordering issues. However, it could leave more room for optimizations in the different broker trees, where it could again be sensible to apply the CI heuristic presented in Chapter 5, for example. Finding a good partitioning into different trees is another interesting problem which depends on the application scenario.

Adaptation and Reconfiguration

Seamless reconfiguration is a prerequisite for self-management and self-optimization which are of increasing importance because of a growing demand for applications that can cope with dynamic scenarios. The middleware layer is a promising place for realizing reconfiguration because it makes reconfiguration transparent for the application developer and provides means for adding

---

[1]http://www.solacesystems.com (last visit: 2008-09-29)

self-management and adaptation features in an application-independent way [67]. Thus, the application developer does not need to be an expert in self-management anymore but can rely on self-management features offered by the middleware.

In this book, the focus was on the reconfiguration of the broker overlay topology for a certain class of routing algorithms. The next step would be to generalize the reconfiguration mechanism for arbitrary correct routing algorithms. Moreover, it would be interesting to not only dynamically reconfigure the topology but also the routing algorithm used. This would open up the possibility to switch between different routing algorithms on the fly, depending on the message flows. First steps in this direction have been undertaken by Bittner and Hinze recently [24].

The integration of reconfiguration and self-optimization in Chapters 4 and 5 already showed that a modular architecture of the publish/subscribe system is a necessary prerequisite in order to be able to implement dynamic reconfigurations. Baldoni et al. [15] and Cugola and Picco [51] already presented flexible architectures that support limited reconfiguration at runtime. However, more research is necessary in this direction regarding, for example, the dynamic growth of the notification service (i.e., starting and stopping brokers on-the-fly), reconfiguring the routing algorithm used, and flow control.

Appendix

A Parameters Used for Topology Generation

BRITE [109] was used to generate Internet-like topologies for the experiments conducted in Section 5.6 and Section 4.4.3. The parameterization of BRITE is documented in Table A.1 on the next page.

Topology Type	2 Level: TOP-DOWN
AS	
HS:	1000
N:	100
LS:	100
Model:	Waxman
Node Placement:	Heavy Tailed
alpha:	0.15
Growth Type:	Incremental
beta:	0.2
Pref. Conn.:	None
m:	2
Route	
HS:	1000
N:	100
LS:	100
Model:	Waxman
Node Placement:	Heavy Tailed
alpha:	0.15
Growth Type:	Incremental
beta:	0.2
Pref. Conn.:	None
m:	2
Top Down	
Edge Connection Model:	Random
Inter BW Distr.:	Heavy Tailed
Max BW:	1024
Min BW:	10
Intra BW Distr.:	Heavy Tailed
Max BW:	1024
Min BW:	10

Table A.1: Parameters used for topology generation with BRITE

B Example Topologies Generated by the Heuristics

Figures B.1, B.2, and B.3 on the following pages show topologies as generated by the different heuristics in Experiment 4 on page 224 with non-uniformly distributed clients. Figure B.4 shows the static random topology generated at the beginning of the experiment. The layout of the depicted topologies is done automatically and does not respect any cost values. The size of the circles that represent a broker illustrate different load values (the bigger the size, the higher the load value). Gray, box-shaped nodes represent brokers without any clients.

The simulation runs were seeded with the same values such that the load values as well as the communication and processing costs are the same in all topologies.

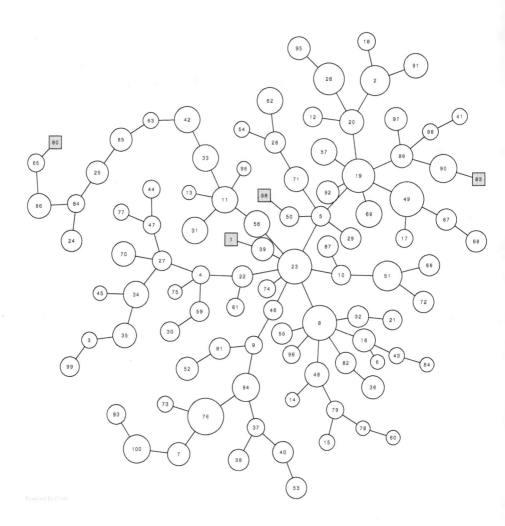

Figure B.1: Topology created by the CI heuristic

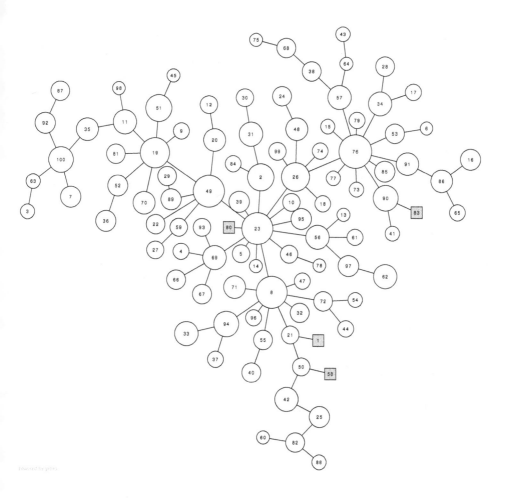

Licensed by press

Figure B.2: Topology created by the ℓMIT heuristic

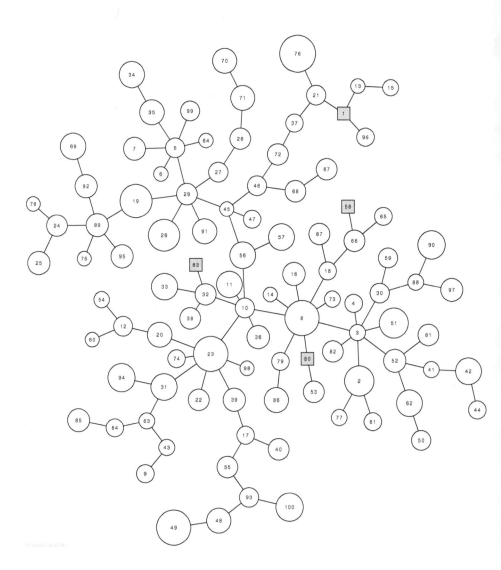

Figure B.3: Topology created by the ℓMST heuristic

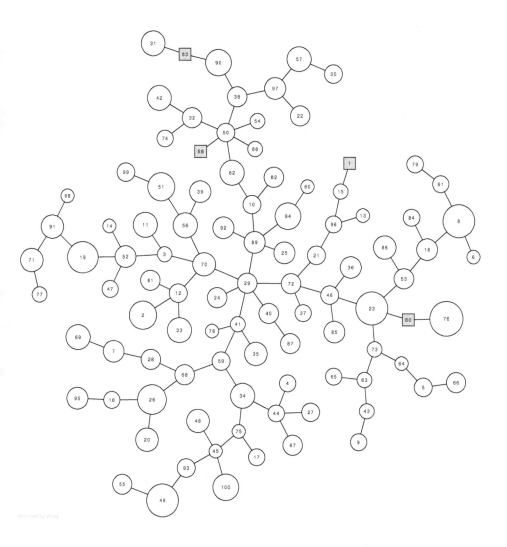

Figure B.4: Random topology created at the beginning of the experiment

Bibliography

[1] ADLER, M., GE, Z., KUROSE, J. F., TOWSLEY, D. F., AND ZABELE, S. Channelization problem in large scale data dissemination. In *Proceedings of the 9th International Conference on Network Protocols (ICNP'01)* (Nov. 2001), IEEE Computer Society, pp. 100–109.

[2] AEKATERINIDIS, I., AND TRIANTAFILLOU, P. PastryStrings: A comprehensive content-based publish/subscribe DHT network. In *Proceedings of the 26th IEEE International Conference on Distributed Computing Systems (ICDCS'06)* (July 2006), H. Ichikawa and M. Raynal, Eds., IEEE, IEEE Computer Society, p. 23.

[3] AFEK, Y., KUTTEN, S., AND YUNG, M. Memory-efficient self stabilizing protocols for general networks. In *Proceedings of the 4th International Workshop on Distributed Algorithms (WDAG'90)* (Sept. 1990), J. van Leeuwen and N. Santoro, Eds., vol. 486 of *Lecture Notes in Computer Science*, Springer-Verlag, pp. 15–28.

[4] AGGARWAL, S., AND KUTTEN, S. Time optimal self-stabilizing spanning tree algorithms. In *Proceedings of the 13th Conference on Foundations of Software Technology and Theoretical Computer Science (FSTTCS'93)* (1993), Springer-Verlag, pp. 400–410.

[5] AGGARWAL, V., FELDMANN, A., AND SCHEIDELER, C. Can ISPs and P2P systems co-operate for improved performance? *ACM SIGCOMM Computer Communications Review (CCR) 37*, 3 (July 2007), 29–40.

[6] AGUILERA, M., STROM, R., STURMAN, D., ASTLEY, M., AND CHANDRA, T. Matching events in a content-based subscription system. In *Proceedings of the 18th ACM Symposium on Principles of Distributed Computing (PODC'99)* (1999), pp. 53–61.

[7] ALTINEL, M., AND FRANKLIN, M. J. Efficient filtering of XML documents for selective dissemination of information. In *Proceedings of 26th International Conference on Very Large Data Bases (VLDB'00)* (Sept. 2000), A. E. Abbadi, M. L. Brodie, S. Chakravarthy, U. Dayal, N. Kamel, G. Schlageter, and K.-Y. Whang, Eds., Morgan Kaufmann, pp. 53–64.

[8] ANCEAUME, E., DATTA, A. K., GRADINARIU, M., SIMON, G., AND VIRGILLITO, A. DPS: Self-* dynamic and reliable content-based publish/subscribe system. Tech. Rep. PI 1665, Institut de Recherche En Informatique Et Systémes Aléatoires (IRISA), Rennes, France, 2004.

[9] ANCEAUME, E., DATTA, A. K., GRADINARIU, M., SIMON, G., AND VIRGILLITO, A. A semantic overlay for self-* peer-to-peer publish/subscribe. In *Proceedings of the 26th IEEE International Conference on Distributed Computing Systems (ICDCS'06)* (July 2006), H. Ichikawa and M. Raynal, Eds., IEEE, IEEE Computer Society, p. 22.

[10] ARORA, A., AND GOUDA, M. Distributed Reset. *IEEE Transactions on Computers 43*, 9 (Sept. 1994), 1026–1038.

[11] ARORA, A., AND GOUDA, M. G. Closure and convergence: A foundation of fault-tolerant computing. *Software Engineering 19*, 11 (1993), 1015–1027.

[12] ARORA, A., AND WANG, Y.-M. Practical self-stabilization for tolerating unanticipated faults in networked systems. Tech. Rep. OSU-CISRC-1/03-TR01, The Ohio State University, Computer Information Systems Department, 2003.

[13] BALDONI, R., BERALDI, R., PIERGIOVANNI, S. T., AND VIRGILLITO, A. On the modelling of publish/subscribe communication systems. *Concurrency and Computation: Practice and Experience 17*, 12 (Oct. 2005), 1471–1495.

[14] BALDONI, R., BERALDI, R., QUEMA, V., QUERZONI, L., AND TUCCI-PIERGIOVANNI, S. TERA: Topic-based event routing for peer-to-peer architectures. In Jacobsen et al. [79], pp. 2–13.

[15] BALDONI, R., BERALDI, R., QUERZONI, L., AND VIRGILLITO, A. A self-organizing crash-resilient topology management system for content-based publish/subscribe. In *Proceedings of the 3rd International Workshop on Distributed Event-Based Systems (DEBS'04)* (May 2004), A. Carzaniga and P. Fenkam, Eds., IEE, pp. 3–8.

[16] BALDONI, R., BERALDI, R., QUERZONI, L., AND VIR-GILLITO, A. Subscription-driven self-organization in content-based publish/subscribe. Tech. rep., DIS, Mar. 2004. http://www.dis.uniroma1.it/~midlab/docs/BBQV04techrep.pdf (last access: 2007-10-21).

[17] BALDONI, R., BERALDI, R., QUERZONI, L., AND VIRGILLITO, A. Efficient publish/subscribe through a self-organizing broker overlay and its application to SIENA. *The Computer Journal 50*, 4 (July 2007), 444–459.

[18] BALDONI, R., MARCHETTI, C., VIRGILLITO, A., AND VITENBERG, R. Content-based publish-subscribe over structured overlay networks. In *Proceedings of the 25th International Conference on Distributed Computing Systems (ICDCS'05)* (June 2005), IEEE, IEEE Computer Society, pp. 437–446.

[19] BANERJEE, S., KOMMAREDDY, C., KAR, K., BHATTACHARJEE, B., AND KHULLER, S. Construction of an efficient overlay multicast infrastructure for real-time applications. In *Proceedings of the 22nd Annual Joint Confer-*

ence of the IEEE Computer and Communications Societies (INFOCOM'03) (Mar./Apr. 2003), vol. 2, IEEE, pp. 1521–1531.

[20] BEHNEL, S., FIEGE, L., AND MÜHL, G. On quality-of-service and publish/subscribe. In *Proceedings of the 5th International Workshop on Distributed Event-Based Systems (DEBS'06)* (July 2006), A. Hinze and J. Pereira, Eds., IEEE Press, p. 20.

[21] BHARAMBE, A. R., RAO, S., AND SESHAN, S. Mercury: A scalable publish-subscribe system for internet games. In *Proceedings of 1st Workshop on Network and System Support for Games (NetGames'02)* (Apr. 2002), ACM Press, pp. 3–9.

[22] BHOLA, S., STROM, R., BAGCHI, S., ZHAO, Y., AND AUERBACH, J. Exactly-once delivery in a content-based publish-subscribe system. In *Proceedings of the International Conference on Dependable Systems and Networks (DSN'02)* (June 2002), J.-C. Fabre and F. Jahanian, Eds., IEEE Press, pp. 7–16.

[23] BIRMAN, K. P., AND JOSEPH, T. A. Reliable communication in the presence of failures. *ACM Transactions on Computer Systems 5*, 1 (1987), 47–76.

[24] BITTNER, S., AND HINZE, A. The arbitrary boolean publish/subscribe model: Making the case. In Jacobsen et al. [79], pp. 226–237.

[25] BLOOM, B. H. Space/time trade-offs in hash coding with allowable errors. *Communications of the ACM 13*, 7 (1970), 422–426.

[26] BORNHÖVD, C., CILIA, M., LIEBIG, C., AND BUCHMANN, A. An infrastructure for meta-auctions. In *Proceedings of the 2nd International Workshop on Advanced Issues of E-Commerce and Web-based Information Systems (WECWIS'00)* (June 2000), IEEE Computer Society, pp. 21–30.

[27] BRICCONI, G., NITTO, E. D., AND TRACANELLA, E. Issues in analyzing the behavior of event dispatching systems. In *Proceedings of the 10th International Workshop on Software Specification and Design (IWSSD'00)* (2000), IEEE Computer Society, pp. 95–103.

[28] BURNS, J. E., GOUDA, M. G., AND MILLER, R. E. Stabilization and pseudo-stabilization. *Distributed Computing 7*, 1 (1993), 35–42.

[29] CAMPAILLA, A., CHAKI, S., CLARKE, E., JHA, S., AND VEITH, H. Efficient filtering in publish-subscribe systems using binary decision diagrams. In *Proceedings of the 23rd International Conference on Software Engineering (ICSE'01)* (Toronto, Canada, May 2001), pp. 443–452.

[30] CAO, F., AND SINGH, J. P. Efficient event routing in content-based publish/subscribe service networks. In *Proceedings of the 23rd Conference of the IEEE Communications Society (INFOCOM'04)* (Mar. 2004), V. O. K. Li, B. Li, and M. Krunz, Eds., IEEE Communications Society.

[31] CAPORUSCIO, M., MARCO, A. D., AND INVERARDI, P. Run-time performance management of the Siena publish/subscribe middleware. In *Proceedings of the 5th International Workshop on Software and Performance (WOSP'05)* (New York, NY, USA, 2005), ACM Press, pp. 65–74.

[32] CARVALHO, N., ARAUJO, F., AND RODRIGUES, L. Scalable QoS-based event routing in publish-subscribe systems. In *Proceedings of the 4th IEEE International Symposium on Network Computing and Applications (NCA'05)* (July 2005), IEEE Computer Society, pp. 101–108.

[33] CARZANIGA, A. *Architectures for an Event Notification Service Scalable to Wide-area Networks*. PhD thesis, Politecnico di Milano, Milano, Italy, Dec. 1998.

[34] CARZANIGA, A., AND HALL, C. P. Content-based communication: a research agenda. In *Proceedings of the 6th International Workshop on Soft-*

ware Engineering and Middleware (SEM'06) (Nov. 2006), E. Wohlstadter, Ed., ACM Press, pp. 2–8. Invited Paper.

[35] CARZANIGA, A., AND WOLF, A. L. Content-based networking: A new communication infrastructure. In *Developing an Infrastructure for Mobile and Wireless Systems (NSF Workshop IMWS 2001)* (Oct. 2001), vol. 2538 of *Lecture Notes in Computer Science (LNCS)*, Springer-Verlag, pp. 59–68.

[36] CARZANIGA, A., AND WOLF, A. L. Forwarding in a content-based network. In *Proceedings of the 2003 Conference on Applications, Technologies, Architectures, and Protocols for Computer Communications (SIGCOMM'03)* (Aug. 2003), A. Feldmann, M. Zitterbart, J. Crowcroft, and D. Wetherall, Eds., ACM Press, pp. 163–174.

[37] CASTELLI, S., COSTA, P., AND PICCO, G. P. Modeling the communication costs of content-based routing: The case of subscription forwarding. In Jacobsen et al. [79], pp. 38–49.

[38] CASTRO, M., DRUSCHEL, P., HU, Y. C., AND ROWSTRON, A. I. T. Topology-aware routing in structured peer-to-peer overlay networks. In *Future Directions in Distributed Computing: Research and Position Papers* (2003), A. Schiper, A. A. Shvartsman, H. Weatherspoon, and B. Y. Zhao, Eds., vol. 2584 of *Lecture Notes in Computer Science (LNCS)*, Springer-Verlag, pp. 103–107.

[39] CASTRO, M., DRUSCHEL, P., KERMARREC, A. M., AND ROWSTRON, A. I. T. Scribe: A large-scale and decentralized application-level multicast infrastructure. *IEEE Journal on Selected Areas in Communications 20*, 8 (Oct. 2002), 1489–1499.

[40] CAYLEY, A. A theorem on trees. *Quart. J. Math 23* (1889), 376–378.

[41] CHAPPELL, D. A. *Enterprise Service Bus*, 1 ed. O'Reilly, June 2004.

[42] CHOCKLER, G., MELAMED, R., TOCK, Y., AND VITENBERG, R. Constructing scalable overlays for pub-sub with many topics. In *Proceedings of the 26th Annual ACM Symposium on Principles of Distributed Computing (PODC'07)* (Aug. 2007), ACM, ACM Press, pp. 109–118.

[43] CHOCKLER, G., MELAMED, R., TOCK, Y., AND VITENBERG, R. SpiderCast: A scalable interest-aware overlay for topic-based pub/sub communication. In Jacobsen et al. [79], pp. 14–25.

[44] CHOI, Y., PARK, K., AND PARK, D. HOMED: A peer-to-peer overlay architecture for large-scale content-based publish/subscribe systems. In *Proceedings of the 3rd International Workshop on Distributed Event-Based Systems (DEBS'04)* (Edinburgh, Scotland, UK, May 2004), A. Carzaniga and P. Fenkam, Eds., IEE, pp. 20–25.

[45] CLARK, D. D. The design philosophy of the DARPA internet protocols. *SIGCOMM Computer Communication Review 25*, 1 (1995), 102–111.

[46] COSTA, P., MIGLIAVACCA, M., PICCO, G. P., AND CUGOLA, G. Introducing reliability in content-based publish-subscribe through epidemic algorithms. In *Proceedings of the 2nd International Workshop on Distributed Event-Based Systems (DEBS'03)* (June 2003), H.-A. Jacobsen, Ed., ACM Press, pp. 1–8.

[47] COSTA, P., MIGLIAVACCA, M., PICCO, G. P., AND CUGOLA, G. Epidemic algorithms for reliable content-based publish-subscribe: An evaluation. In *Proceedings of the 24th International Conference on Distributed Computing Systems (ICDCS'04)* (Mar. 2004), IEEE Computer Society, pp. 552–561.

[48] COSTA, P., AND PICCO, G. P. Semi-probabilistic content-based publish-subscribe. In *Proceedings of the 25th International Conference on Distributed*

Computing Systems (ICDCS'05) (June 2005), IEEE, IEEE Computer Society, pp. 575–585.

[49] COSTA, P., PICCO, G. P., AND ROSSETTO, S. Publish-subscribe on sensor networks: A semi-probabilistic approach. In *Proceedings of the 2nd IEEE International Conference on Mobile Ad-hoc and Sensor Systems (MASS'05)* (Nov. 2005), IEEE, p. 10.

[50] CUGOLA, G., FREY, D., MURPHY, A. L., AND PICCO, G. P. Minimizing the reconfiguration overhead in content-based publish-subscribe. In *Proceedings of the 2004 ACM Symposium on Applied Computing (SAC'04)* (New York, NY, USA, 2004), ACM Press, pp. 1134–1140.

[51] CUGOLA, G., AND PICCO, G. P. REDS: A reconfigurable dispatching system. In *Proceedings of the 6th International Workshop on Software Engineering and Middleware (SEM'06)* (Nov. 2006), E. Wohlstadter, Ed., ACM Press, pp. 9–16.

[52] CUGOLA, G., PICCO, G. P., AND MURPHY, A. L. Towards dynamic reconfiguration of distributed publish-subscribe middleware. In *Proceedings of the 3rd International Workshop on Software Engineering and Middleware (SEM'02)* (2002), W. Emmerich, A. Coen-Porisini, and A. van der Hoek, Eds., vol. 2596 of *Lecture Notes in Computer Science (LNCS)*, Springer-Verlag, pp. 187–202.

[53] DABEK, F., ZELDOVICH, N., KAASHOEK, F., MAZIÈRES, D., AND MORRIS, R. Event-driven programming for robust software. In *Proceedings of the 10th Workshop on ACM SIGOPS European Workshop* (New York, NY, USA, 2002), ACM Press, pp. 186–189.

[54] DATTA, A. K., GRADINARIU, M., RAYNAL, M., AND SIMON, G. Anonymous publish/subscribe in P2P networks. In *Proceedings of the International Parallel and Distributed Processing Symposium (IPDPS'03)* (Apr. 2003),

M. Cosnard, A. Gottlieb, and J. Dongarra, Eds., IEEE Computer Society, pp. 74–81.

[55] DEMERS, A., GREENE, D., HAUSER, C., IRISH, W., AND LARSON, J. Epidemic algorithms for replicated database maintenance. In *Proceedings of the 6th Annual ACM Symposium on Principles of Distributed Computing (PODC'87)* (1987), ACM Press, pp. 1–12.

[56] DIJKSTRA, E. W. Self-stabilizing systems in spite of distributed control. *Communications of the ACM 17*, 11 (1974), 643–644.

[57] DOLEV, S. *Self-Stabilization.* MIT Press, Mar. 2000.

[58] DOLEV, S., AND HERMAN, T. Superstabilizing protocols for dynamic distributed systems. *Chicago Journal of Theoretical Computer Science 4* (Dec. 1997). Special Issue on Self-Stabilization.

[59] DOLEV, S., ISRAELI, A., AND MORAN, S. Self-stabilization of dynamic systems assuming only read/write atomicity. In *Proceedings of the 9th Annual ACM Symposium on Principles of Distributed Computing (PODC'90)* (New York, NY, USA, 1990), ACM Press, pp. 103–117.

[60] DOLEV, S., AND KAT, R. I. HyperTree for self-stabilizing peer-to-peer systems. In *Proceedings of the 3rd IEEE International Symposium on Network Computing and Applications (NCA'04)* (Washington, DC, USA, 2004), IEEE, pp. 25–32.

[61] EUGSTER, P., FELBER, P., GUERRAOUI, R., AND KERMARREC, A.-M. The many faces of publish/subscribe. *ACM Computing Surveys 35*, 2 (2003), 114–131.

[62] FENNER, B., HANDLEY, M., HOLBROOK, H., AND KOUVELAS, I. Protocol independent multicast – sparse mode (PIM-SM): Protocol specification (revised). RFC 4601 (Proposed Standard), Aug. 2006.

[63] FIEGE, L., MÜHL, G., AND GÄRTNER, F. C. Modular event-based systems. *The Knowledge Engineering Review 17*, 4 (2003), 359–388.

[64] FREY, D. *Publish Subscribe on Large-Scale Dynamic Topologies: Routing and Overlay Management*. PhD thesis, Politecnico Di Milano, 2006.

[65] GARCÍA, J., JAEGER, M. A., MÜHL, G., AND BORRELL, J. Decoupling components of an attack prevention system using publish/subscribe. In *Proceedings of the 2005 IFIP Conference on Intelligence in Communication Systems (INTELLCOMM'04)* (Oct. 2005), R. Glitho, A. Karmouch, and S. Pierre, Eds., vol. 190, IFIP, Springer-Verlag, pp. 87–97.

[66] GÄRTNER, F. C. A survey of self-stabilizing spanning-tree construction algorithms. Tech. Rep. 200338, Swiss Federal Institute of Technology (EPFL), School of Computer and Communication Sciences, June 2003.

[67] GEIHS, K. Middleware challenges ahead. *Computer 34*, 6 (June 2001), 24–31.

[68] GHOSH, S., GUPTA, A., HERMAN, T., AND PEMMARAJU, S. Fault-containing self-stabilizing algorithms. In *Proceedings of the 15th Annual ACM Symposium of Distributed Computing (PODC'96)* (May 1996), ACM Press, pp. 45–54.

[69] GHOSH, S., AND PEMMARAJU, S. Trade-offs in fault-containing self-stabilization. In *Proceedings of the 3rd Workshop on Self-Stabilizing Systems (WSS'97)* (1997), S. Ghosh and T. Herman, Eds., Charleton University Press, pp. 157–169.

[70] GUPTA, A., SAHIN, O. D., AGRAWAL, D., AND ABBADI, A. E. Meghdoot: Content-based publish/subscribe over P2P networks. In *Proceedings of the 5th ACM/IFIP/USENIX International Conference on Middleware (MIDDLEWARE'04)* (Oct. 2004), H.-A. Jacobsen, Ed., Springer-Verlag, pp. 254–273.

[71] GUPTA, S. K. S., AND SRIMANI, P. K. Self-stabilizing multicast protocols for ad hoc networks. *Journal of Parallel and Distributed Computing 63*, 1 (2003), 87–96.

[72] HANSEN, S., AND FOSSUM, T. Events not equal to GUIs. *SIGCSE Bulletin 36*, 1 (2004), 378–381.

[73] HERMAN, T. Superstabilizing mutual exclusion. Tech. Rep. TR 97-04, University of Iowa, Iowa, USA, 1997.

[74] HERMAN, T. Models of self-stabilization and sensor networks. In *Proceedings of the 5th International Workshop on Distributed Computing (IWDC'03)* (Dec. 2003), S. R. Das and S. K. Das, Eds., vol. 2918 of *Lecture Notes in Computer Science (LNCS)*, Springer-Verlag, pp. 205–214.

[75] HERRMANN, K. *Self-Organizing Infrastructures for Ambient Services*. PhD thesis, Berlin University of Technology, 2006.

[76] HOHPE, G. Programmieren ohne Stack: ereignis-getriebene Architekturen. *OBJEKTspektrum 02* (Feb. 2006), 18–24. In German.

[77] HU, T. Optimum communication spanning trees. *SIAM Journal on Computing 3*, 3 (1974), 188–195.

[78] INTANAGONWIWAT, C., GOVINDAN, R., ESTRIN, D., HEIDEMANN, J., AND SILVA, F. Directed diffusion for wireless sensor networking. *ACM/IEEE Transactions on Networking 11*, 1 (Feb. 2002), 2–16.

[79] JACOBSEN, H.-A., MÜHL, G., AND JAEGER, M. A., Eds. *Proceedings of the Inaugural Conference on Distributed Event-Based Systems (DEBS'07)* (New York, NY, USA, June 2007), ACM Press.

[80] JAEGER, M. A. Self-Organizing Publish/Subscribe. In *Proceedings of the 2nd International Doctoral Symposium on Middleware (MDS'05)* (New York, NY, USA, Nov. 2005), E. Curry, Ed., ACM Digital Library, pp. 1–5.

[81] JAEGER, M. A. Self-Organizing Publish/Subscribe. *IEEE Distributed Systems Online 7*, 2 (Feb. 2006). art. no. 0602-o2003.

[82] JAEGER, M. A., AND MÜHL, G. Stochastic analysis and comparison of self-stabilizing routing algorithms for publish/subscribe systems. In *Proceedings of the 13th IEEE/ACM International Symposium on Modeling, Analysis and Simulation of Computer and Telecommunication Systems (MASCOTS'05)* (Sept. 2005), G. F. Riley, R. Fujimoto, and H. Karatza, Eds., IEEE Press, pp. 471–479.

[83] JAEGER, M. A., MÜHL, G., WERNER, M., AND PARZYJEGLA, H. Reconfiguring self-stabilizing publish/subscribe systems. In *Proceedings of the 17th IFIP/IEEE International Workshop on Distributed Systems: Operations and Management (DSOM'06)* (Oct. 2006), R. State, S. van Meer, D. O'Sullivan, and T. Pfeifer, Eds., vol. 4269 of *Lecture Notes in Computer Science (LNCS)*, Springer-Verlag, pp. 233–238.

[84] JAEGER, M. A., PARZYJEGLA, H., MÜHL, G., AND HERRMANN, K. Self-organizing broker topologies for publish/subscribe systems. In *Proceedings of the 22nd Annual ACM Symposium on Applied Computing (SAC'07)* (Mar. 2007), L. M. Liebrock, Ed., ACM, pp. 543–550.

[85] JAYARAM, M., AND VARGHESE, G. Crash failures can drive protocols to arbitrary states. In *Proceedings of the 15th Annual ACM Symposium on Principles of Distributed Computing (PODC'96)* (New York, NY, USA, 1996), ACM Press, pp. 247–256.

[86] JIANG, T., AND LI, Q. A self-stabilizing distributed multicast algorithm for mobile ad-hoc networks. In *Proceedings of the The 4th International Conference on Computer and Information Technology (CIT'04)* (2004), IEEE Computer Society, pp. 499–502.

[87] JIN, Y., AND STROM, R. Relational subscription middleware for internet-scale publish-subscribe. In *Proceedings of the 2nd International Workshop on Distributed Event-Based Systems (DEBS'03)* (June 2003), H.-A. Jacobsen, Ed., ACM Press, pp. 1–8.

[88] JOHNSON, D. S., LENSTRA, J. K., AND KAN, A. H. G. R. The complexity of the network design problem. *Networks 8* (1978), 279–285.

[89] JUNGINGER, M. O., AND LEE, Y. A self-organizing publish/subscribe middleware for dynamic peer-to-peer networks. *IEEE Network 18*, 1 (2004), 38–43.

[90] KAKUGAWA, H., AND YAMASHITA, M. A dynamic reconfiguration tolerant self-stabilizing token circulation algorithm in ad-hoc networks. In *Proceedings of the 8th International Conference on Principles of Distributed Systems (OPODIS'04)* (Dec. 2004), T. Higashino, Ed., vol. 3544 of *Lecture Notes in Computer Science (LNCS)*, Springer-Verlag, pp. 256–266. Revised Selected Papers.

[91] KARGER, D., LEHMAN, E., LEIGHTON, T., PANIGRAHY, R., LEVINE, M., AND LEWIN, D. Consistent hashing and random trees: Distributed caching protocols for relieving hot spots on the world wide web. In *Proceedings of the 29th Annual ACM Symposium on Theory of Computing (STOC'97)* (New York, NY, USA, 1997), ACM Press, pp. 654–663.

[92] KARGER, D. R., AND RUHL, M. Finding nearest neighbors in growth-restricted metrics. In *Proceedings of the 34th Annual ACM Symposium on Theory of Computing (STOC'02)* (New York, NY, USA, 2002), ACM Press, pp. 741–750.

[93] KEPHART, J. O., AND CHESS, D. M. The vision of autonomic computing. *IEEE Computer 36*, 1 (Jan. 2003), 41–50.

[94] KLEINROCK, L. *Queuing Systems; Theory*, vol. 1. John Wiley and Sons, New York, 1975.

[95] KLEINROCK, L. *Queuing Systems; Computer Applications*, vol. 2. John Wiley and Sons, New York, 1976.

[96] KRUSKAL, J. B. On the shortest spanning subtree of a graph and the traveling salesman problem. *Proceedings of the American Mathematical Society 7*, 1 (Feb. 1956), 48–50.

[97] KUMAR S.D., M., AND BELLUR, U. A distributed algorithm for underlay aware and available overlay formation in event broker networks for publish/subscribe systems. In *Proceedings of the 27th International Conference on Distributed Computing Systems Workshops (ICDCSW'07)* (June 2007), pp. 69–69.

[98] KWON, M., AND FAHMY, S. Topology-aware overlay networks for group communication. In *Proceedings of the 12th International Workshop on Network and Operating Systems Support for Digital Audio and Video (NOSS-DAV'02)* (New York, NY, USA, 2002), ACM Press, pp. 127–136.

[99] LAMPORT, L. Time, clocks, and the ordering of events in a distributed system. *Communications of the ACM 21*, 7 (1978), 558–565.

[100] LAMPORT, L. A new approach to proving the correctness of multiprocess programs. *ACM Transactions on Programming Languages and Systems (TOPLAS) 1*, 1 (1979), 84–97.

[101] LENDARIS, G. G. On the definition of self-organizing systems. *Proceedings of the IEEE 52*, 3 (Mar. 1964), 324–325.

[102] LI, Y., AND BOUCHEBABA, Y. A new genetic algorithm for the optimal communication spanning tree problem. In *Proceedings of the 4th European Conference on Artificial Evolution (AE'99)* (Nov. 2000), C. Fonlupt, J.-K.

Hao, E. Lutton, E. M. A. Ronald, and M. Schoenauer, Eds., vol. 1829 of *Lecture Notes in Computer Science (LNCS)*, Springer-Verlag, pp. 162–173. Selected Papers.

[103] LIU, H., AND JACOBSEN, H.-A. A-TOPSS – a publish/subscribe system supporting approximate matching. In *Proceedings of 28th International Conference on Very Large Data Bases (VLDB'02)* (Aug. 2002), Morgan Kaufmann, pp. 1107–1110.

[104] LUMEZANU, C., BHOLA, S., AND ASTLEY, M. Utility optimization for event-driven distributed infrastructures. In *Proceedings of the 26th IEEE International Conference on Distributed Computing Systems (ICDCS'06)* (July 2006), H. Ichikawa and M. Raynal, Eds., IEEE, IEEE Computer Society, pp. 24–24.

[105] LUMEZANU, C., SPRING, N., AND BHATTACHARJEE, B. Decentralized message ordering for publish/subscribe systems. In *ACM/IFIP/USENIX 7th International Middleware Conference* (Nov. 2006), M. van Steen and M. Henning, Eds., vol. 4290 of *Lecture Notes in Computer Science (LNCS)*, Springer-Verlag, pp. 162–179.

[106] LWIN, C. H., MOHANTY, H., AND GHOSH, R. K. Causal ordering in event notification service systems for mobile users. In *Proceedings of the International Conference on Information Technology: Coding and Computing (ITCC'04)* (Apr. 2004), vol. 2, IEEE Computer Society, pp. 735–740.

[107] LWIN, C. H., MOHANTY, H., GHOSH, R. K., AND CHAKRABORTY, G. Resilient dissemination of events in a large-scale event notification service system. In *Proceedings of the IEEE International Conference on e-Technology, e-Commerce and e-Service (EEE'05)* (Mar. 2005), pp. 502–507.

[108] MALKIN, G. RIP version 2. RFC 2453 (Standard), Nov. 1998.

[109] MEDINA, A., LAKHINA, A., MATTA, I., AND BYERS, J. BRITE: An approach to universal topology generation. In *Proceedings of the International Workshop on Modeling, Analysis and Simulation of Computer and Telecommunications Systems (MASCOTS'01)* (Aug. 2001), IEEE Computer Society, pp. 346–353.

[110] MERZ, P., AND WOLF, S. TreeOpt: Self-organizing, evolving P2P overlay topologies based on spanning trees. In *KiVS 2007 Workshop: Selbstorganisierende, Adaptive, Kontextsensitive verteilte Systeme (SAKS'07)* (Mar. 2007), T. Braun, G. Carle, and B. Stiller, Eds., ITG/GI, VDE Verlag, pp. 231–242.

[111] MICHAEL, L., NEJDL, W., PAPAPETROU, O., AND SIBERSKI, W. Improving distributed join efficiency with extended bloom filter operations. In *Proceedings of the 21st International Conference on Advanced Information Networking and Applications (AINA'07)* (May 2007), pp. 187–194.

[112] MIGLIAVACCA, M., AND CUGOLA, G. Adapting publish-subscribe routing to traffic demands. In Jacobsen et al. [79], pp. 91–96. Short paper.

[113] MITZENMACHER, M. Network applications of bloom filters: A survey. *Internet Mathematics 1*, 4 (2006), 485–509.

[114] MITZENMACHER, M., AND UPFAHL, E. *Probability and Computing: Randomized Algorithms and Probabilistic Analysis*. Cambridge University Press, Cambridge, UK, 2005.

[115] MÜHL, G. *Large-Scale Content-Based Publish/Subscribe Systems*. PhD thesis, Darmstadt University of Technology, Sept. 2002.

[116] MÜHL, G., FIEGE, L., AND PIETZUCH, P. R. *Distributed Event-Based Systems*. Springer-Verlag, Aug. 2006.

[117] MÜHL, G., JAEGER, M. A., HERRMANN, K., WEIS, T., FIEGE, L., AND ULBRICH, A. Self-stabilizing publish/subscribe systems: Algorithms and evaluation. In *Proceedings of the 11th International Conference on Parallel Processing (Euro-Par 2005)* (2005), J. C. Cunha and P. D. Medeiros, Eds., vol. 3648 of *Lecture Notes in Computer Science (LNCS)*, Springer-Verlag, pp. 664–674.

[118] MÜHL, G., WERNER, M., JAEGER, M. A., HERRMANN, K., AND PARZY-JEGLA, H. On the definitions of self-managing and self-organizing systems. In *KiVS 2007 Workshop: Selbstorganisierende, Adaptive, Kontextsensitive verteilte Systeme (SAKS'07)* (Mar. 2007), T. Braun, G. Carle, and B. Stiller, Eds., ITG/GI, VDE Verlag, pp. 291–301.

[119] NAICKEN, S., LIVINGSTON, B., BASU, A., RODHETBHAI, S., WAKEMAN, I., AND CHALMERS, D. The state of peer-to-peer simulators and simulations. *SIGCOMM Computer Communication Review 37*, 2 (2007), 95–98.

[120] OPYRCHAL, L., ASTLEY, M., AUERBACH, J., BANAVAR, G., STROM, R., AND STURMAN, D. Exploiting IP multicast in content-based publish-subscribe systems. In *Proceedings of the IFIP/ACM International Conference on Distributed Systems Platforms (MIDDLEWARE'00)* (Apr. 2000), J. S. Sventek and G. Coulson, Eds., vol. 1795 of *Lecture Notes in Computer Science (LNCS)*, Springer-Verlag, pp. 185–207.

[121] PAPAEMMANOUIL, O., AHMAD, Y., ÇETINTEMEL, U., JANNOTTI, J., AND YILDIRIM, Y. Extensible optimization in overlay dissemination trees. In *Proceedings of the ACM SIGMOD International Conference on Management of Data (SIGMOD'06)* (June 2006), S. Chaudhuri, V. Hristidis, and N. Polyzotis, Eds., ACM, pp. 611–622.

[122] PARZYJEGLA, H., MÜHL, G., AND JAEGER, M. A. Reconfiguring publish/subscribe overlay topologies. In *Proceedings of the 5th International*

Workshop on Distributed Event-Based Systems (DEBS'06) (July 2006), A. Hinze and J. Pereira, Eds., IEEE Press, p. 29.

[123] PATON, N. W., AND DÍAZ, O. Active database systems. *ACM Computing Surveys 31*, 1 (1999), 63–103.

[124] PELEG, D., AND RESHEF, E. Deterministic polylog approximation for minimum communication spanning trees. In *Proceedings of the 25th International Colloquium on Automata, Languages and Programming (ICALP'98)* (July 1998), K. G. Larsen, S. Skyum, and G. Winskel, Eds., vol. 1443 of *Lecture Notes in Computer Science (LNCS)*, Springer-Verlag, pp. 670–681.

[125] PETROVIC, M., LIU, H., AND JACOBSEN, H.-A. CMS-ToPSS: Efficient dissemination of RSS documents. In *Proceedings of the International Conference on Very Large Databases (VLDB'05)* (2005), K. Böhm, C. S. Jensen, L. M. Haas, M. L. Kersten, P.-Å. Larson, and B. C. Ooi, Eds., ACM, pp. 1279–1282.

[126] PICCO, G. P., CUGOLA, G., AND MURPHY, A. L. Efficient content-based event dispatching in the presence of topological reconfiguration. In *Proceedings of the 23rd International Conference on Distributed Computing Systems (ICDCS'03)* (May 2003), IEEE Computer Society, pp. 234–243.

[127] PIETZUCH, P., SHNEIDMAN, J., LEDLIE, J., WELSH, M., SELTZER, M., AND ROUSSOPOULOS, M. Evaluating DHT-based service placement for stream-based overlays. In *Proceedings of the 4th International Workshop on Peer-to-Peer Systems (IPTPS'05)* (Feb. 2005), M. Castro and R. van Renesse, Eds., vol. 3640 of *Lecture Notes in Computer Science (LNCS)*, Springer-Verlag, pp. 275–286.

[128] PIETZUCH, P. R. *Hermes: A Scalable Event-Based Middleware.* PhD thesis, Computer Laboratory, Queens' College, University of Cambridge, Feb. 2004.

[129] PIETZUCH, P. R., AND BACON, J. Peer-to-peer overlay broker networks in an event-based middleware. In *Proceedings of the 2nd International Workshop on Distributed Event-Based Systems (DEBS'03)* (June 2003), H.-A. Jacobsen, Ed., ACM Press, pp. 1–8.

[130] PNUELI, A. The temporal semantics of concurrent programs. *Theoretical Computer Science 13* (1981), 45–60.

[131] PRIM, R. C. Shortest connection networks and some generalizations. *Bell System Technical Journal 36* (1957), 1389–1401.

[132] QUERZONI, L. *Techniques for efficient event routing.* PhD thesis, Università di Roma "La Sapienzia", Rome, Italy, Mar. 2007.

[133] RATNASAMY, S., FRANCIS, P., HANDLEY, M., KARP, R., AND SCHENKER, S. A scalable content-addressable network. In *Proceedings of the 2001 Conference on Applications, Technologies, Architectures, and Protocols for Computer Communications (SIGCOMM'01)* (New York, NY, USA, 2001), ACM Press, pp. 161–172.

[134] RIABOV, A., LIU, Z., WOLF, J. L., YU, P. S., AND ZHANG, L. Clustering algorithms for content-based publication-subscription systems. In *Proceedings of the 22nd International Conference on Distributed Computing Systems (ICDCS'02)* (July 2002), IEEE Computer Society, pp. 133–142.

[135] ROSE, I., MURTY, R., PIETZUCH, P., LEDLIE, J., ROUSSOPOULOS, M., AND WELSH, M. Cobra: Content-based filtering and aggregation of blogs and RSS feeds. In *Proceedings of the 4th USENIX Symposium on Networked Systems Design & Implementation* (Apr. 2007), pp. 29–42.

[136] ROWSTRON, A., AND DRUSCHEL, P. Pastry: Scalable, decentralized object location and routing for large-scale peer-to-peer systems. In *Proceeedings of the IFIP/ACM International Conference on Distributed Systems Platforms*

(MIDDLEWARE'01) (Nov. 2001), R. Guerraoui, Ed., vol. 2218 of *Lecture Notes in Computer Science (LNCS)*, Springer-Verlag, pp. 329–350.

[137] SACHS, K., AND KOUNEV, S. Workload scenario for SPECjms "Supermarket Supply Chain". Tech. Rep. DVS06-1, SPEC OSG Java Subcommittee, 2006.

[138] SCHNEIDER, M. Self-stabilization. *ACM Computing Surveys (CSUR) 25*, 1 (1993), 45–67.

[139] SCHWARTZ, J. Who needs hackers? Online, Sept. 2007. http://nytimes.com/2007/09/12/technology/techspecial/12threat.html, Last access: 2007-09-15.

[140] SEETHARAMAN, S., AND AMMAR, M. On the interaction between dynamic routing in native and overlay layers. In *Proceedings of the 25th IEEE International Conference on Computer Communications (INFOCOM'06)* (Apr. 2006), IEEE, pp. 1–12.

[141] SHARMA, P., ESTRIN, D., FLOYD, S., AND JACOBSON, V. Scalable timers for soft state protocols. In *Proceedings of the 26th IEEE International Conference on Computer Communications (INFOCOM'97)* (Apr. 1997), pp. 222–229.

[142] SHEN, Z., AND TIRTHAPURA, S. Self-stabilizing routing in publish-subscribe systems. In *Proceedings of the 3rd International Workshop on Distributed Event-Based Systems (DEBS'04)* (May 2004), A. Carzaniga and P. Fenkam, Eds., IEE, pp. 92–97.

[143] SHEN, Z., AND TIRTHAPURA, S. Self-stabilizing routing in publish-subscribe systems. Tech. Rep. TR-2004-04-4, Iowa State University, Iowa, USA, Apr. 2004.

[144] STOICA, I., MORRIS, R., KARGER, D., KAASHOEK, M. F., AND BAL-
AKRISHNAN, H. Chord: A scalable peer-to-peer lookup service for Inter-
net applications. In *Proceedings of the 2001 Conference on Applications,
Technologies, Architectures, and Protocols for Computer Communications
(SIGCOMM'01)* (New York, NY, USA, 2001), ACM Press, pp. 149–160.

[145] STROM, R. Gryphon: An information flow based approach to message brokering. In *Proceedings of the 9th International Symposium on
Software Reliability Engineering (ISSRE'98)* (Nov. 1998). Fast abstract
(http://www.research.ibm.com/distributedmessaging/paper5.html, last access 2007-10-
26).

[146] STROM, R. Fault-tolerance in the SMILE stateful publish-subscribe system.
In *Proceedings of the 3rd International Workshop on Distributed Event-Based
Systems (DEBS'04)* (May 2004), A. Carzaniga and P. Fenkam, Eds., IEE,
pp. 98–103.

[147] TANNER, A., AND MÜHL, G. A formalisation of message-complete pub-
lish/subscribe systems. Tech. Rep. Rote Reihe 2004/11, Berlin University of
Technology, Oct. 2004. Brief Announcement at the 18th Annual Conference
on Distributed Computing (DISC'04).

[148] TERPSTRA, W. W., BEHNEL, S., FIEGE, L., ZEIDLER, A., AND BUCH-
MANN, A. P. A peer-to-peer approach to content-based publish/subscribe.
In *Proceedings of the 2nd International Workshop on Distributed Event-
Based Systems (DEBS'03)* (June 2003), H.-A. Jacobsen, Ed., ACM Press,
pp. 1–8.

[149] TRIANTAFILLOU, P., AND AEKATERINIDIS, I. Content-based publish-
subscribe over structured P2P. In *Proceedings of the 3rd International
Workshop on Distributed Event-Based Systems (DEBS'04)* (May 2004),
A. Carzaniga and P. Fenkam, Eds., IEE, pp. 104–109.

[150] TRIANTAFILLOU, P., AND ECONOMIDES, A. Subscription summaries for scalability and efficiency in publish/subscribe systems. In *Proceedings of the 1st International Workshop on Distributed Event-Based Systems (DEBS'02)* (Vienna, Austria, 2002), J. Bacon, L. Fiege, R. Guerraoui, H.-A. Jacobsen, and G. Mühl, Eds., IEEE Press, pp. 619–624. Published as part of the ICDCS '02 Workshop Proceedings.

[151] VARGHESE, G., AND JAYARAM, M. The fault span of crash failures. *Journal of the ACM (JACM) 47*, 2 (2000), 244–293.

[152] VOULGARIS, S. *Epidemic-Based Self-Organization in Peer-to-Peer Systems.* PhD thesis, Vrije Universiteit, Amsterdam, The Netherlands, 2006.

[153] VOULGARIS, S., RIVIERE, E., KERMARREC, A.-M., AND VAN STEEN, M. Sub-2-Sub: Self-organizing content-based publish subscribe for dynamic large scale collaborative networks, 2006. No published proceedings (online available at: http://people.inf.ethz.ch/spyros/papers/iptps.2006.pdf, last access 2007-10-26).

[154] WEIS, T., PARZYJEGLA, H., JAEGER, M. A., AND MÜHL, G. Self-organizing and self-stabilizing role assignment in sensor/actuator networks. In *Proceedings of the 8th International Symposium on Distributed Objects and Applications (DOA'06)* (Oct. 2006), R. Meersman and Z. Tari, Eds., vol. 4276 of *Lecture Notes in Computer Science (LNCS)*, Springer-Verlag, pp. 1807–1824.

[155] WILLEMS, J. C. Paradigms and puzzles in the theory of dynamical systems. *IEEE Transactions on Automatic Control 36*, 3 (Mar. 1991), 259–294.

[156] WONG, T., KATZ, R., AND MCCANNE, S. An evaluation of preference clustering in large-scale multicast applications. In *Proceedings of the 19th Annual Joint Conference of the IEEE Computer and Communications Societies (INFOCOM'00)* (Mar. 2000), vol. 2, pp. 451–460.

[157] WU, B. Y., CHAO, K.-M., AND TANG, C. Y. Approximation algorithms for some optimum communication spanning tree problems. In *Proceedings of the 9th International Symposium of Algorithms and Computation (ISAAC'98)* (Dec. 1998), K.-Y. Chwa and O. H. Ibarra, Eds., vol. 1533 of *Lecture Notes in Computer Science (LNCS)*, Springer-Verlag, pp. 407–416.

[158] WU, B. Y., LANCIA, G., BAFNA, V., CHAO, K.-M., RAVI, R., AND TANG, C. Y. A polynomial time approximation scheme for minimum routing cost spanning trees. In *Proceedings of the 9th Annual ACM-SIAM Symposium on Discrete Algorithms (SODA'98)* (Jan. 1998), SIAM Press, pp. 21–32.

[159] XU, Z., AND SRIMANI, P. K. Self-stabilizing publish/subscribe protocol for P2P networks. In *Proceedings of the 7th International Workshop on Distributed Computing (IWDC'05)* (Dec. 2005), A. Pal, A. D. Kshemkalyani, R. Kumar, and A. Gupta, Eds., vol. 3741 of *Lecture Notes in Computer Science (LNCS)*, Springer-Verlag, pp. 129–140.

[160] YAN, T., AND GARCIA-MOLINA, H. Index structures for selective dissemination of information under the Boolean model. *ACM Transactions on Database Systems (TODS) 19*, 2 (June 1994), 332–364.

[161] YONEKI, E., AND BACON, J. An adaptive approach to content-based subscription in mobile ad hoc network. In *Proceedings of the 2nd IEEE Annual Conference on Pervasive Computing and Communications, Workshop on Mobile Peer-to-Peer Computing (MP2P'04)* (Mar. 2004), IEEE Computer Society, pp. 92–97.

[162] YU, H., ESTRIN, D., AND GOVINDAN, R. A hierarchical proxy architecture for Internet-scale event services. In *Proceedings of the 8th IEEE International Workshop on Enabling Technologies: Infrastructure for Collaborative Enterprises (WET ICE '99)* (1999), pp. 78–83.

[163] ZADEH, L. A. On the definition of adaptivity. *Proceedings of the IEEE 51*, 3 (1963), 469–470.

[164] ZHAO, Y., BHOLA, S., AND STURMAN, D. C. Subscription propagation and content-based routing with delivery guarantees. In *Proceedings of the 19th International Conference on Distributed Computing (DISC'05)* (Sept. 2005), P. Fraigniaud, Ed., vol. 3724 of *Lecture Notes in Computer Science (LNCS)*, Springer-Verlag, pp. 501–502.

[165] ZHAO, Y., STURMAN, D., AND BHOLA, S. Subscription propagation in highly-available publish/subscribe middleware. In *Proceedings of the 6th International Middleware Conference (MIDDLEWARE'05)* (Nov. 2005), G. Alonso, Ed., vol. 3790 of *Lecture Notes in Computer Science (LNCS)*, Springer-Verlag, pp. 274–293.

[166] ZHOU, Y., OOI, B. C., TAN, K.-L., AND YU, F. Adaptive reorganization of coherency-preserving dissemination tree for streaming data. In *Proceedings of the 22nd International Conference on Data Engineering (ICDE'06)* (Apr. 2006), L. Liu, A. Reuter, K.-Y. Whang, and J. Zhang, Eds., IEEE Computer Society, p. 55.

[167] ZHU, Y., LI, B., AND GUO, J. Multicast with network coding in application-layer overlay networks. *IEEE Journal on Selected Areas in Communications 22*, 1 (Jan. 2004), 107–120.

Index

VDM
Verlag
Dr. Müller

Wissenschaftlicher Buchverlag bietet

kostenfreie

Publikation

von

wissenschaftlichen Arbeiten

Diplomarbeiten, Magisterarbeiten, Master und Bachelor Theses
sowie Dissertationen, Habilitationen und wissenschaftliche Monographien

Sie verfügen über eine wissenschaftliche Abschlußarbeit zu aktuellen oder zeitlosen
Fragestellungen, die hohen inhaltlichen und formalen Ansprüchen genügt,
und haben **Interesse an einer honorarvergüteten Publikation**?

Dann senden Sie bitte erste Informationen über Ihre Arbeit per Email
an info@vdm-verlag.de. Unser Außenlektorat meldet sich umgehend bei Ihnen.

VDM Verlag Dr. Müller Aktiengesellschaft & Co. KG
Dudweiler Landstraße 125a
D - 66123 Saarbrücken

www.vdm-verlag.de

www.ingramcontent.com/pod-product-compliance
Lightning Source LLC
LaVergne TN
LVHW022302060326
83290ZLV00020B/3219